Leszek Wroński

**Reichenbach's Paradise**

**Constructing the Realm of Probabilistic Common "Causes"**

Leszek Wroński

# Reichenbach's Paradise

## Constructing the Realm of Probabilistic Common "Causes"

---

Managing Editor: Anna Michalska

Language Editor: Sara Tavares

DE GRUYTER
OPEN

Published by De Gruyter Open Ltd, Warsaw/Berlin



ISBN 978-3-11-037270-0
e-ISBN 978-3-11-037271-7

**Bibliographic information published by the Deutsche Nationalbibliothek**
The Deutsche Nationalbibliothek lists this publication in the Deutsche Nationalbibliografie; detailed bibliographic data are available in the Internet at http://dnb.dnb.de.

Managing Editor: Anna Michalska
Language Editor: Sara Tavares

www.degruyteropen.com

Cover illustration: © IStock/VikaSuh

To my parents

# Contents

## List of Figures



## List of Tables

## Acknowledgments

First and foremost, my thanks go to Professor Tomasz Placek, the best PhD supervisor in any bundle of branching space-times.

My interest in the subject sparked during the '08 Summer School in Probabilistic Causality at the CEU University in Budapest. I would like to thank Professor Miklós Rédei for introducing me to the problems discussed in chapter 5.

I owe a great deal to Michał Marczyk, with whom I extensively worked on the issues of causal closedness of probability spaces. Most of the material in chapters 5 and 6 originates from our joint research; also, theorem 1 from chapter 2 is our joint result.

I am grateful to the faculty and PhD students of the Epistemology Department at the Institute of Philosophy of the Jagiellonian University for insightful comments after my talks regarding the issues covered in this book. I would also like to thank Jakub Byszewski, Lech Duraj, Michał Skrzypczak, Michał Staromiejski and Andrzej Wroński for helpful discussions on mathematical topics.

On the side of the Publisher I would like to thank Anna Michalska, the Managing Editor for Philosophy, for supervising the preparation of this book. The comments of the Language Editor, whom I only know as "Sara" (and who has not seen this paragraph), were invaluable; in one case they lead to a reformulation of a section discussing open formal problems. I would also like to express great thanks to the two anonymous Reviewers whose detailed input enabled me to—I hope—improve the book. Special thanks are due to the Reviewer who suggested changing the title.

I would also like to thank the friendly internet community of RPG Codex for some moments of inspiration.

## Relation with papers in print or previously published

Most of the mathematical results herein stem from my PhD dissertation (Wroński, 2010) which in reality forms the bulk of this book. For various reasons most of them will only now appear in article form. Let me then say here (I will give more detail when I reach the specific cases) that some of the theorems and proofs in this essay correspond to those found in Wroński and Marczyk (2013), Marczyk and Wroński (2014) and Wroński and Marczyk (2010).

## Introduction

Philosophical concepts change through history, with new notions becoming intertwined with old ones, while familiar principles acquire new formulations. With ongoing progress in mathematics, some philosophical ideas obtain a formal dimension, which may shed light on unresolved issues. This has happened e.g. with the notion of cause, in which case theories of probabilistic causality added rigour and clarity to former discussions. We should of course keep in mind Aristotle's well-known point that "it is the mark of an educated man to look for precision in each class of things just so far as the nature of the subject admits"[1]. Some people make this the basis of their opinion that the higher the level of precision, the lower the potential philosophical insight. I agree there is a risk of losing track of philosophy with a capital "$\Phi$" among formal details, but in my opinion it is fascinating that one can apply formal tools to analyse some aspects of a philosophical notion. If these tools are mathematical, the endeavour should simply be called "mathematical philosophy". But there is the other side of the medal: some mathematically oriented philosophers believe "non-formal" philosophy to be inferior in some way—I do not share this opinion. One can be a philosopher in a variety of ways. Contrary to what some "formal" philosophers may think, non-formal philosophy is not "easier". Solving an open formal problem may be not more difficult than arriving at a good argument in an informal context, especially when the notion of a problem and its openness is vague at best. On the other hand, only after attaining informal insights do we know which formal questions to ask.

Still, most of the insights I would like to believe this book to have are of a formal nature. I will very briefly point at some non-formal directions in which one could continue dealing with the topic, but I will not pursue them. I would rather like to investigate in detail the formal aspects of the notion of common cause—which has, since the work by Hans Reichenbach, achieved an explication in at least partially mathematically tractable terms of events and probabilities—and, in particular, the "common cause principle".

In one of its most basic and informal shapes, the common cause principle states that any surprising correlation between two factors which are believed not to directly influence one another is due to their (possibly hidden) common cause. In the history of philosophy it is easy to find examples of similar reasoning; one needs to look no further than the mind-body problem. There is a truly astonishing correlation between our thoughts of the "I want to wave my hand" sort and the movements-of-our-hands of the waving sort. A venerable solution to this quandary is that of invoking God as the common cause (which was the road taken e.g. by Malebranche).

We can perhaps look for similar causal intuitions in Mill's *System of Logic*. From the fifth Canon of Induction it follows that a concomitant variation in two phenomena of which none is a cause of the other is a sign of a connection between the two by "some fact of causation". Mill begins his exposition of the Canon by referring to the case in which this fact is the phenomena being two effects of a common cause (Vol. I, Book III, Chapter VIII of Mill (1868)). Bertrand Russell is on a similar track when he writes of "identity of structure" leading to "the assumption of a common causal origin" (Russell (2009), p. 409). We, however, will be concerned with an idea which possesses a probabilistic formulation. It was introduced, in the form of a general principle, by Hans Reichenbach in his posthumously published book *The Direction of Time*. The central

1 *Nicomachean Ethics*, bk. 1, ch. 3; translation by W.D. Ross.

notion of the principle in Reichenbach's formulation, and of the current essay, is that of screening off: two correlated events are screened off by a third event if conditioning on the third event makes them probabilistically independent. Reichenbach's principle marks also the beginning of a new field of philosophy, namely, that of "probabilistic causality".

The main results of this work are presented in chapters 5 and 6. For the most part, the current essay can be seen as an effort at checking how far one can go with the purely statistical notions revolving around Reichenbach's idea of common cause. In short, the answer is "surprisingly far"; in some classes of probability spaces all correlations between "interesting"[2] events possess explanations of such sort; moreover, it will be shown that any probability space (i.e. any space describing some experiment) can be extended to a more "fine-grained" space in which for any correlation there exists a Reichenbach-style common cause. Chapters 5 and 6 contain mathematical results concerning these issues, regarding not only the exact notion used by Reichenbach but also some of its modern variations. The results will lend themselves to opposing interpretations: on the one hand, they are positive answers to some open mathematical problems, stating the virtually universal existence of explanatory events satisfying philosophically motivated mathematical conditions; on the other, though, whether these events will be of genuine explanatory value in general, whether they will correspond to some concrete phenomena we could use in subsequent explanations is highly doubtful.

In chapter 1, after some mathematical preliminaries, we study the various formulations of the principle which might be said to stem from the original idea of Reichenbach. We also examine a few of the most salient counterarguments, which undermine at least some of the formulations. Chapter 2 is of a formal nature, dealing with various probabilistic notions which can be thought of as generalizations of Reichenbach's concept of common cause. The next chapter concerns the relationship between the idea of common causal explanation and the Bell inequalities. In chapter 4 we briefly present the formulation of Reichenbach's principle which can be found in the field of representing causal structures by means of directed acyclic graphs.

**Funding information**

Support from:

- the Polish Ministry of Science and Higher Education PhD Grant N N101 131936;
- the Polish Ministry of Science and Higher Education research grant 668/N-RNP-ESF/2010/0 (head: Tomasz Placek);
- and the Foundation for Polish Science "START" Fellowship

is gratefully acknowledged.

2 E.g. "logically independent", this will be formally defined in chapter 5.

# 1 The Principle of the Common Cause: Its Shapes and Content

## 1.1 Probability: the Basics

Before we state the various forms of the Principle, some of which will be of a formal nature, a few definitions are in order.

**Definition 1 [Probability space]** A *probability space* is a triple $\langle \Omega, \mathcal{F}, P \rangle$ such that:
- $\Omega$ is a nonempty set;
- $\mathcal{F}$ is a nonempty family of subsets of $\Omega$ which is closed under complement and countable union;
- $P$ is a function from $\mathcal{F}$ to $[0, 1] \subset \mathbb{R}$ such that
  - $P(\Omega) = 1$;
  - $P$ is countably additive: for a countable family $\mathcal{G}$ of pairwise disjoint members of $\mathcal{F}$, $P(\cup \mathcal{G}) = \sum_{A \in \mathcal{G}} P(A)$.

In the context of a probability space $\langle \Omega, \mathcal{F}, P \rangle$, $\Omega$ is called the *sample space*, $\mathcal{F}$ is called the *event algebra*, and $P$ is called the *probability function* (or *measure*). The members of $\mathcal{F}$ are called *events*.

The above definition captures the content of the concept of a *classical* probability space. In one of the chapters to come we will also discuss *non-classical* spaces, but let us postpone their definition till then. Also, in a later chapter we will treat probability spaces as *pairs* consisting of a Boolean algebra and a measure defined on it; this is because we will be speaking mostly about finite structures for which any Boolean algebra of subsets is of course complete with regard to the operations of complement and countable union (and *vice versa*, any such family is a Boolean algebra). In general, though, it may be that a Boolean algebra of subsets of a given set is incomplete with respect to (w.r.t.) the operation of countable union.

The complement of an event $A$, $\mathcal{F} \setminus A$, will be written as "$A^{\perp}$". If it is evident from the context that $A$ and $B$ are events, we will sometimes write "$P(AB)$" instead of "$P(A \wedge B)$" or "$P(A \cap B)$" for "the probability that both $A$ and $B$ occur".

Every event $B \in \mathcal{F}$ such that $P(B) \neq 0$ determines a measure $P_B$ on the same event algebra: namely, for any $A \in \mathcal{F}$, $P_B(A) := \frac{P(AB)}{P(B)}$. We define the *conditional probability of A given B* to be equal to $P_B(A)$; to refer to it, we will almost exclusively use the traditional notation "$P(A \mid B)$". If $P(B) = 0$, we take $P(A \mid B)$ to be undefined.

We will now define the concept of a random variable. Our main reference is Feller (1968), but the formulation of some definitions is inspired by Forster (1988). Since in what follows we do not use continuous random variables, we can omit the usual measure-theoretic definitions; in fact, we will only need random variables with a finite number of possible values. This is why by "random variable" we will mean what is traditionally referred to as "finite-valued random variable". In what follows, $X^{-1}(v)$ is the preimage of $v$ under the function $X$.

**Definition 2 [Random variable]** Let $\langle \Omega, \mathcal{F}, P \rangle$ be a probability space. Let $V$ be a finite subset of $\mathbb{R}$. A *random variable* on $\Omega$ is a function $X : \Omega \to V$ such that

$$\forall v \in V \quad X^{-1}(v) \in \mathcal{F}.$$

If $|V| = 2$, $X$ is called a *binary random variable.*

Thus every random variable determines a set of events directly tied to its values. The expression "$P(X = v)$", "probability that the random variable $X$ takes the value $v$", is to be understood as "$P(X^{-1}(v))$"; this is straightforwardly generalized for any subset of $V$, so that for any $V' \subseteq V$, $P(X \in V') = \sum_{v \in V'} P(X = v)$.

Though random variables as defined above are real-valued functions, on some occasions it might of course be useful to think of them as functions with values of a different type, e.g. expressions "yes" or "no". In numerical contexts below we will always treat binary random variables as if they assume values 0 and 1.

It is immediate that a random variable $X : \Omega \to V$ can be thought of as a method of dividing the sample space $\Omega$ into at most $|V|$ pieces—the preimages of the members of $V$. There are two intuitive and important ways of thinking about this, depending on our view of the sample space.

First, $\Omega$ can be considered to consist of all possible outcomes of an experiment—for example, if the experiment is a single toss of a six-sided die, then $\Omega = \{1, 2, 3, 4, 5, 6\}$. A random variable may correspond to a *feature* which some outcomes possess; for example, if $X(1) = X(3) = X(5) =$ "*yes*", and $X(2) = X(4) = X(6) =$ "*no*", then the feature might be "being odd", and "$P(X =$ "*yes*"$)$" is to be interpreted as "the probability that the outcome of the toss is odd".

On the other hand, sometimes $\Omega$ is to be viewed not as a set of outcomes of an experiment, but rather as the *population* on which an experiment is conducted. Suppose a group of people is tested for a virus. $\Omega$ will then consist of the test subjects, and $P(X =$ "*yes*"$)$ will mean "the probability that a randomly chosen test subject has the virus".

Notice also the close correspondence of events and binary random variables. An event $A$ is a subset of the sample space; we can construct a binary random variable so that the preimage of "*yes*" is $A$ and the preimage of "*no*" is $A^{\perp}$. Similarly, any binary random variable gives rise (by way of the preimages of its values) to two events, $A$ and $A^{\perp}$.

The concept of "correlation" usually concerns random variables, but in the literature around the Principle of the Common Cause it has frequently been defined for event, too.[1] (Usually no probability spaces are defined and the notion of event is an intuitive one, frequently that of a space-time region.) Since in the course of this work we will mostly be talking about events, and not random variables, we shall continue that practice and begin with the simpler concept.

**Definition 3 [Correlation (events)]** Let $\langle \Omega, \mathcal{F}, P \rangle$ be a probability space and let $A$, $B \in \mathcal{F}$. We say that $A$ and $B$ are:
- *positively correlated*, or just *correlated*, whenever $P(AB) > P(A)P(B)$;
- *negatively correlated*, or *anti-correlated*, whenever $P(AB) < P(A)P(B)$;
- *uncorrelated*, or *(probabilistically) independent*, whenever $P(AB) = P(A)P(B)$.

To define correlation for random variables, we need the notion of covariance and, for that, the notion of expected value.

**Definition 4 [Correlation (variables)]** Let $\langle \Omega, \mathcal{F}, P \rangle$ be a probability space and $X : \Omega \rightarrow V$, $Y : \Omega \rightarrow W$ be random variables on $\Omega$.
- The *expected value* of $X$ is defined as

$$E(X) := \sum_{v \in V} v \cdot P(v).^{2}$$

- The *covariance* of $X$ and $Y$ is defined as

$$Cov(X, Y) := E\Big( (X - E(X)) \cdot (Y - E(Y)) \Big) = E(XY) - E(X)E(Y).$$

- The *correlation coefficient* of $X$ and $Y$ is defined as

$$\rho(X, Y) = \frac{Cov(X, Y)}{\sqrt{Cov(X, X)} \cdot \sqrt{Cov(Y, Y)}}.$$

  The correlation coefficient is not defined if $Cov(X, X)$ or $Cov(Y, Y)$ is equal to 0. We will say the variables $X$ and $Y$ are *correlated* whenever $\rho(X, Y) > 0$.

In the context of the Principle of the Common Cause, what demands explanation is a *correlation* between events or a *dependence* between random variables. As for the

---

1 The relation between the two notions is described e.g. in Forster (1988).

2 A mathematically informed Reader may well expect that we add here something like "whenever the right-hand side of the expression converges; we then say that $X$ has a *finite expectation*"; similarly, that we should prefix the next two clauses of the definition with "Suppose $X$ and $Y$ have finite expectations". But notice that by definition we are dealing here with finite random variables only and so no question of convergence arises.

latter, some recent authors (e.g. Reiss (2007)) say simply that "Two variables $X$ and $Y$ are probabilistically dependent just in case $P(XY) \neq P(X)P(Y)$." Let us expand this into a definition.

**Definition 5 [Dependence (variables)]** Let $\langle \Omega, \mathcal{F}, P \rangle$ be a probability space and let $X : \Omega \to V$ and $Y : \Omega \to W$ be two random variables on $\Omega$. $X$ and $Y$ are *dependent* if

$$\exists V' \subseteq V \, \exists W' \subseteq W : \; P(X \in V' \wedge Y \in W') \neq P(X \in V')P(Y \in W').$$

Note that for random variables the concepts of independence and noncorrelation diverge. If two variables are independent, their correlation coefficient is 0, but not always *vice versa*; for examples see Feller (1968), p. 236. Still, to restate the above, a non-zero correlation coefficient means the variables are dependent.

For binary variables $X$ and $Y$ their covariance is obviously equal to

$$P(X = 1 \wedge Y = 1) - P(X = 1)P(Y = 1).$$

This explains why definition 3 can be seen as a special case of definition 4; events are correlated whenever their corresponding binary variables are, and *vice versa*.

*A note about the pair notation.* When we want to refer to the ordered pair containing $A$ and $B$, we use the expression "$\langle A, B \rangle$". Sometimes what we need is just the two-element set $\{A, B\}$; we frequently also refer to such sets as "pairs". The meaning is always evident given the context.

### 1.1.1 Screening Off

Perhaps the most important notion concerning the idea of common causes is one of *screening off*.

**Definition 6 [Screening off]** Assume a probability space $\langle \Omega, \mathcal{F}, P \rangle$ is given. Let $A, B \in \mathcal{F}$. An event $C$ is said to be a *screener-off* for the pair $\{A, B\}$ if

$$P(AB \mid C) = P(A \mid C)P(B \mid C). \tag{1.1}$$

In the case where $A$ and $B$ are correlated we also say that $C$ *screens off* the correlation.

If $C$ is a screener-off for $\{A, B\}$, we will also frequently say that $C$ "screens off $A$ from $B$" and *vice versa*. Another way of putting the fact is saying that $C$ renders $A$ and $B$ conditionally probabilistically independent. Observe that the screening off condition 1.1 is equivalent to the following:

$$P(A \mid BC) = P(A \mid C) \tag{1.2}$$

provided all the probabilities are defined.

**Definition 7 [Statistical relevance]** Let $\langle \Omega, \mathcal{F}, P\rangle$ be a probability space. Let $A, B \in \mathcal{F}$. We say that an event $C \in \mathcal{F}$ is *positively statistically relevant* for $A$ if $P(A|C) > P(A|C^{\perp})$. We say that a family of events $\{C_i\}$ is *statistically relevant* for $A$ and $B$ if whenever $i \neq j$

$$\left(P(A \mid C_i) - P(A \mid C_j)\right)\left(P(B \mid C_i) - P(B \mid C_j)\right) > 0.$$

Notice that $P(A|C) > P(A|C^{\perp})$ is equivalent to $P(A|C) > P(A)$, if all the probabilities are defined.

### 1.1.2 Observing Probabilities and Correlations

We now have the requisite definitions of probability and related concepts. But how do we observe probabilities "in the world"? If probabilities are to be "limiting frequencies", as one notable interpretation would have it, then we have a problem, since, as beings capable of only finitely many operations, we naturally observe only relative frequencies in finite samples. We can only pose hypotheses about probabilities—but these hypotheses may be well-grounded, thanks to the law of large numbers (see e.g. Feller (1968), p. 243).

If the probability of our experiment ending with a particular outcome is $\phi$, the more we repeat the experiment, the closer should the observed relative frequency of the outcome come to $\phi$. If the probability is unknown, the question regarding the number of repetitions needing to be conducted for us to be able to offer a reliable hypothesis regarding it is subtle, and the answer to it depends on how reliable we require the hypothesis to be. These issues are treated extensively e.g. in Blalock (1979). The technical details will not be of interest to us; the important thing is that no reliable information about probabilities of particular events (and so, *a fortiori*, about their correlation, as well as probability distributions and correlation of random variables) can be gathered from a small experimental sample.

It will be worthwhile to reiterate this point in an analysis of single occurrences of events which we find unexpected or surprising. Suppose, for instance, that someone rolled two fair six-sided dice on a flat table and ended up with two sixes. Why are we (a bit) surprised? Is the result improbable? That particular combination (a six on the first die, a six on the second die) is improbable to exactly the same degree (1/36) as any other possible combination; so the reason for the surprise must be something different. And perhaps it is two-fold:

1. the sum of the results (12) is maximally different from the expected value (7); and perhaps we implicitly compare the probability of rolling it (1/36) with the probability of rolling 11 or more (1/12), 10 or more (1/6) and so on;

2. all throws end up with the same score, which is quite an improbable event (1/6) compared to the alternative, which we implicitly expect to occur.

If $X_i$ is the event "die number $i$ ends up with a 6", then $P(X_1X_2) = \frac{1}{36}$; but just from the occurrence of that particular event we by no means infer that $P(X_1X_2) > P(X_1)P(X_2)$. The reason for the fact that a single occurrence of an improbable coincidence, being a conjunction of other events, startles us, is not that we perceive it as evidence of a yet unsuspected correlation. Otherwise we would always have to accuse lucky dice players, or have pity on unlucky ones, for playing with unfair dice.

This is not to say that a proponent of the frequentist interpretation of probability necessarily cannot speak in any way about probabilities of single events. Reichenbach himself would be an example to the contrary—his way of ascribing probabilities to single events is described in section 72 of Reichenbach (1949). Even though he states on p. 375 that single-case probability is a "pseudo-concept", he develops a way of thinking about the single case as "the limit of the [reference] classes becoming gradually narrower and narrower" (*ibid.*). However, on his account single-case probabilities are, in contrast to "regular" probabilities, dependent on the state of our knowledge; on the whole, he regards "the statement about the probability of the single case, not as having a meaning of its own, but as an elliptic mode of speech" (*ibid.*, p. 376-377). In any case, a frequentist should not in general let a single occurrence of an event influence his or hers beliefs regarding the probabilities inherent in a given situation.

A different issue is whether the data we are analyzing originates from any sort of probabilistic set-up; whether it is appropriate to consider any underlying probabilities at all. If e.g. some parts of the experiment are influenced by human choice, is it wise to consider the probability of a person choosing a particular option? Cartwright (1999) holds the view that no statements regarding probabilities in the world are true *simpliciter*, but in fact may only be true *ceteris paribus*; they need to arise in the context of a "probability machine", a fixed arrangement of components with stable capacities giving rise to regular behaviour.

> "We can make sense of the probability of drawing two red balls in a row from an urn of a certain composition with replacement; but we cannot make sense of the probability of six percent inflation in the United Kingdom next year without an implicit reference to a specific social and institutional structure that will serve as the chance set-up that generates this probability" (Cartwright (1999), p. 175).[3]

The chance set-ups may be of various kinds: "the stochastic process is the world line of the persisting object (*a die, a socio-economic structure*)" (Reiss (2007), emphasis

3 Chapter 7.4 of Cartwright's book contains a detailed description of a probability machine in the context of probabilistic claims about causality made by Salmon (1971).

ours). With no additional information, though, it is unwise to expect a set of data, and the derived relative frequencies of events, as indicative of probability.

## 1.2 The Plurality of the Principles

The literature on the Principle (henceforth referred to as "PCC") abounds in dissenting opinions regarding its validity. It is "false" (Arntzenius (1992)). It is "non-falsifiable" (Hofer-Szabó et al. (2000)). It is a "fallible epistemic principle" (Reiss (2007)). Lastly, it is "derivable from the second law of thermodynamics[4]" (Reichenbach (1971)). Since each of the above is well-argued for, and since there is such a plethora of views, the subject clearly must be something different in every case. The Principle some authors are arguing against is not always the same Principle their opponents promote.

The multiplicity of forms of the PCC has already been discussed in the literature in e.g. Berkovitz (2000) and sections 3.4 – 3.5 of Placek (2000), but we shall initially take an approach different from those displayed by these texts. Berkovitz analyzes how the prospects of the Principle depend on which concepts of correlation (between types or tokens) and causation are employed. Placek differentiates various versions of the Principle on the basis of the mathematical constitution of the common cause—whether it is a single event or an $n$-tuple of events—and whether it is to explain a single correlation or more. While we will also discuss these important matters later on, right now we propose to consider a gradual process of infusing an initially sketchy and informal Principle with formal content.

Throughout the process we will move from purely "informal" principles to purely "formal" ones. The former may arouse deep intuitions and interesting, yet usually inconclusive discussions; the latter can be formally proved or disproved, but one may doubt their relevance to philosophy, or, in the case an antipathy to all things formal is displayed, to anything interesting at all. This is perhaps the usual case when philosophy meets mathematics: the more formal your considerations, the bigger risk of losing track of significant philosophical content. That said, I have a predilection for formal philosophy, which will perhaps be mostly visible in chapter 5; I find it heartening for a philosopher to be able to prove something from time to time. It would be ideal if an interesting and sound philosophical argumentation could be at least partly based on mathematical proofs.

A side note: probability is a relatively new tool for philosophy. Perhaps a big role in its introduction to philosophy was played by Hans Reichenbach's 1956 book *The Direction of Time* (to which we refer throughout this essay as "Reichenbach (1971)"), where the PCC was first formulated. Subsequently probability has been widely used by researchers in the field of so called "probabilistic causality". To this day, most philoso-

4 Admittedly, only with an additional assumption. See section 1.3.1.

phers writing about probability usually simply use expressions like "$P(A)$" in contexts in which they would normally say "the probability that $A$ occurs", without defining any probability spaces. There is a drawback: the notion of event is foggy. The Reader cannot be sure what qualifies as an event and what does not; he or she is expected to rely on his or hers intuitions. We will see an example where this can result in an unfortunate misunderstanding (see p. 21). I believe that philosophy would benefit from every author defining their probability spaces explicitly, at the cost of their texts becoming perhaps a bit more "dry" and the process of writing them getting more unwieldy.

We will not cite any proponents of the Principles listed below, because it seems almost every participant in the discussion uses a Principle which is in at least one small respect different from most of the others.

**PCC 1** Suppose there is a correlation between two events, which are not directly causally related. Then there exists a common cause of the correlated events.

Notice that no views on the nature of causality are included in the above formulation. While it is difficult to find authors who would openly advocate this view, some arguments offered against "the" PCC (or "Reichenbach's" PCC)—most notably Sober-style examples we will discuss in section 1.4.2—actually negate PCC 1, since the probabilistic description of the allegedly existing common cause is largely irrelevant to the argument.

**PCC 2** Some correlations demand explanation. Of these, some demand explanation by means of a common cause. In each such case there exists a common cause of the correlated events, which renders them conditionally probabilistically independent.

There are two additions in comparison to PCC 1. First, a qualification is added that perhaps only some (not all) correlations stand in need of common causal explanation; some authors use the word "improbable" to describe them. Second, a probabilistic ingredient is added: the postulated common cause of the correlated events should screen them off.

**PCC 3** Let $\langle \Omega, \mathfrak{F}, P \rangle$ be a probability space. For any $A, B \in \mathfrak{F}$ (such that $\langle A, B \rangle$ belongs to a relation of independence $L_{ind}$), if $P(AB) > P(A)P(B)$, then there exists an event $C \in \mathfrak{F}$ (different from both $A$ and $B$) such that

$$\begin{aligned} P(AB \mid C) &= P(A \mid C)P(B \mid C); \\ P(AB \mid C^{\perp}) &= P(A \mid C^{\perp})P(B \mid C^{\perp}); \\ P(A \mid C) &> P(A \mid C^{\perp}); \\ P(B \mid C) &> P(B \mid C^{\perp}). \end{aligned}$$

This version of the Principle is of a formal nature. The word "cause" is nowhere to be seen; it can be of course introduced, by *defining* a common cause for $A$ and $B$ as an event meeting the four requirements above. (We assume this definition for the remainder of this section.) PCC 3 is actually meant to possess two variants: with or without the first expression in parentheses. Frequently a relation of independence is introduced; it is usually *at least* logical independence (so that e.g. the correlation between "heads up" and "tails down" will not stand in need of an explanation in terms of a common cause), and perhaps ideally it is supposed also to cover direct causal independence. However, if $L_{ind}$ is just logical independence, then PCC 3 is simply *false*, as it is easy to find examples of spaces with correlations between logically independent events, for which no event meeting the requirements above exists (see e.g. Hofer-Szabó et al. (2000), p. 91). It is also highly unlikely that "fixing" the relation of independence so that it includes *less* pairs than the relation of purely logical independence will alleviate this difficulty and make the Principle generally plausible. However, an interesting question is: in which classes of probability spaces and for which relations of independence does the Principle hold? We will discuss these issues at length in chapter 5.

To state the last form of the Principle we need to define extension of probability spaces.

**Definition 8 [Extension]** Let $\mathfrak{A} = \langle \Omega, \mathcal{F}, P \rangle$, be a probability space. A space $\mathfrak{A}' = \langle \Omega', \mathcal{F}', P' \rangle$ is called an *extension* of $\mathfrak{A}$ if there is a Boolean algebra embedding $h : \mathcal{F} \to \mathcal{F}'$ which preserves the measure, that is, $\forall A \in \mathcal{F},\ P'(h(A)) = P(A)$.

**PCC 4** Let $\mathfrak{A} = \langle \Omega, \mathcal{F}, P \rangle$ be a probability space. Suppose that $A, B \in \mathcal{F}$ (such that $\langle A, B \rangle$ belongs to a relation of independence $L_{ind}$) are correlated, but there exists *no* $C \in \mathcal{F}$ (different from both $A$ and $B$) such that

$$\begin{aligned} P(AB \mid C) &= P(A \mid C)P(B \mid C); \\ P(AB \mid C^{\perp}) &= P(A \mid C^{\perp})P(B \mid C^{\perp}); \\ P(A \mid C) &> P(A \mid C^{\perp}); \\ P(B \mid C) &> P(B \mid C^{\perp}). \end{aligned}$$

Then there exists a space $\mathfrak{A}' = \langle \Omega', \mathcal{F}', P' \rangle$ such that $\mathfrak{A}'$ is an extension of $\mathfrak{A}$ by means of a homomorphism $h$ and there exists an event $C' \in \mathcal{F}'$ such that

$$\begin{aligned} P'(h(A)h(B) \mid C') &= P'(h(A) \mid C')P'(h(B) \mid C'); \\ P'(h(A)h(B) \mid C'^{\perp}) &= P'(h(A) \mid C'^{\perp})P'(h(B) \mid C'^{\perp}); \\ P'(h(A) \mid C') &> P'(h(A) \mid C'^{\perp}); \\ P'(h(B) \mid C') &> P'(h(B) \mid C'^{\perp}). \end{aligned}$$

As we have already said, there are numerous counterexamples to PCC 3, which is a statement postulating, for each correlation in a given space, a common cause *in the same space*. PCC 4 is, however, more subtle. Suppose we observe an unexpected correlation during an experiment, but the probability space we have chosen to operate within lacks common causes for the correlated events. But perhaps the choice of the space was unfortunate; perhaps we have not taken some factors into account and a different, more “fine-grained” space, compatible with the observations to the same extent as the original one, provides an explanation for the correlation in terms of a common cause? In other words, can the original space be extended to a space possessing a common cause for the yet unexplained correlation? And in general, is it possible to extend a given probability space to one containing common causes for *all* correlations? Perhaps surprisingly, the answer to both questions is “yes”. We will deal with these matters extensively in chapter 6.

We have already mentioned that the place in which the PCC was introduced was Reichenbach’s *Direction of Time*. Subsequently, regardless of the version of the Principle they are concerned with, many authors credit Reichenbach with the original idea. Some of them (e.g. Hoover (2003), p. 527) content themselves with the following quotation: “*If an improbable coincidence has occurred, there must exist a common cause*”[5]. In the next section we will try to convince the Reader that such a selective quotation misses a few important facets of Reichenbach’s view of the Principle.

## 1.3 What Reichenbach Wrote

Reichenbach’s 1956 book is frequently taken to contain an important metaphysical view of probabilistic causality (see e.g. Williamson (2009)). The main object of the book, however, is to analyze the possibilities of defining time direction by means of causal relations. Part IV discusses the case of macrostatistics and it is there, in chapter 19, where the Principle of the Common Cause originally appears.

Throughout his book Reichenbach frequently writes about probability (his formulas will be put here in modern notation), however, he did not choose to adopt the Kolmogorovian concepts of “event algebra” and “probability space”, which were then slowly gaining recognition. The choice was undoubtedly motivated by the fact that he already had his own von Mises-style theory of probability, developed earlier in Reichenbach (1949) (originally issued in German in 1935). It is important to note at the beginning of this section that, for Reichenbach, “the term “probability” is always assumed to mean the limit of a relative frequency” (Reichenbach (1971), p. 123). Therefore the question of probability of an event regarded in isolation of any sequence of its possible occurrences or non-occurrences should be meaningless. To use a popu-

5 Reichenbach (1971), p. 157.

lar philosophical term, for Reichenbach there should be no such things as single-case probabilities. (But remember the quote referring to an "elliptic mode of speech", p. 6.)

The guiding idea behind Reichenbach's Principle, and the source of—as we will see—an important argument for one of Reichenbach's theses is that "the improbable should be explained in terms of causes, not in terms of effects" (Reichenbach (1971), p. 157); the short version of the Principle of the Common Cause quoted at the end of the previous section comes right at the end of the same paragraph. Let us quote the first examples with which Reichenbach's illustrates his Principle (all quotes from *ibid.*, p. 157):

- "Suppose that lightning starts a brush fire, and that a strong wind blows and spreads the fire, which is thus turned into a major disaster. The coincidence of fire and wind has here a common effect, the burning over of a wide area. But when we ask why this coincidence occurred, we do not refer to the common effect, but look for a common cause. The thunderstorm that produced the lightning also produced the wind, and the improbable coincidence is thus explained."
- "Suppose both lamps in a room go out suddenly. We regard it as improbable that by chance both bulbs burned out at the same time, and look for a burned-out fuse or some other interruption of the common power supply. The improbable coincidence is thus explained as the product of a common cause."
- "Or suppose several actors in a stage play fall ill, showing symptoms of food poisoning. We assume that the poisoned food stems from the same source—for instance, that it was contained in a common meal—and thus look for an explanation of the coincidence in terms of a common cause."

Keeping in mind the concept of probability quoted above, up to this point it would hardly seem surprising that the Principle—

**Reichenbach's PCC—the "coincidence" formulation:** "If an improbable coincidence has occurred, there must exist a common cause"

—makes no mention of probability save for the word "improbable" in the antecedent. Reichenbach quickly injects his Principle with more probabilistic content, though. First, he admits that "chance coincidences, of course, are not impossible", since the bulbs may simply burn out at the same moment etc. Therefore, in such cases the existence of a common cause is "not absolutely certain, but only *probable*" (*ibid.*, emphasis ours), with the probability increasing with the number of repeated coincidences. (Let us just note that the concept of probability implicit here seems to be decidedly epistemic—the more repeated coincidences we observe, the more strongly we should believe in the existence of a common cause—and thus hard to reconcile with the earlier definition.) The author offers another two examples supporting the Principle (*ibid.*, p. 158):

- "Suppose two geysers which are not far apart spout irregularly, but throw up their columns of water always at the same time. The existence of a subterranean connection of the two geysers with a common reservoir of hot water is then practically certain."
- "The fact that measuring instruments such as barometers always show the same indication if they are not too far apart, is a consequence of the existence of a common cause—here, the air pressure."

We are then advised to "treat the principle of the common cause as a statistical problem" (*ibid.*). In Reichenbach's view this means that we should assume events $A$ and $B$ have been observed to occur frequently, which enables us to consider probabilities $P(A)$, $P(B)$ and $P(AB)$. The relationship between two (improbably) simultaneously occurring events and both their common cause and effect is depicted in terms of forks seen in figure 1.1.

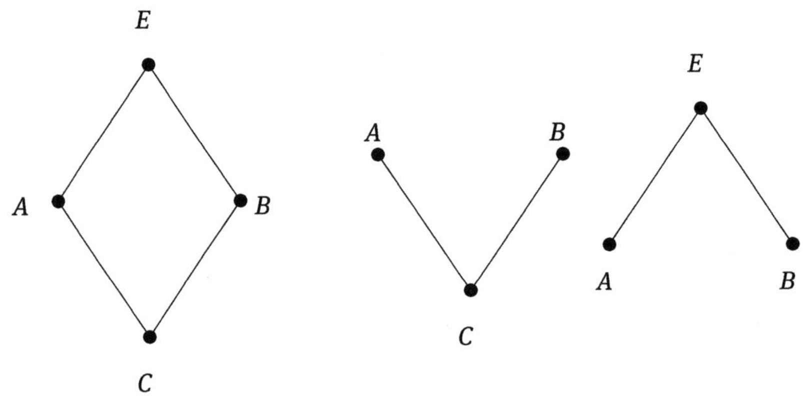


**Figure 1.1:** A double fork, a fork open towards the future, a fork open towards the past. "C" denotes a cause, "E" denotes an effect.

Reichenbach claims the forks depict *statistical* relationships between the events. However, in his examples cited above there always is some physical process behind each arrow on the diagram.

The coincidence of events $A$ and $B$ has, for Reichenbach, "a probability exceeding that of a chance coincidence" (*ibid.*, p. 159) precisely when the two events are correlated in terms of our definition 3. Suppose, then, the events *are* correlated. We "assume that there exists a common cause $C$. If there is more than one possible kind of common cause, $C$ may represent the disjunction of these causes" (*ibid.*). An important assumption is now that the fork $ACB$ satisfies exactly the statistical requirements listed above in the formulation of PCC 3:

$$P(AB \mid C) = P(A \mid C)P(B \mid C); \tag{1.3}$$

$$P(AB \mid C^{\perp}) = P(A \mid C^{\perp})P(B \mid C^{\perp}); \tag{1.4}$$

$$P(A \mid C) > P(A \mid C^{\perp}); \tag{1.5}$$

$$P(B \mid C) > P(B \mid C^{\perp}). \tag{1.6}$$

Namely, both $C$ and $C^{\perp}$ should screen off $A$ from $B$, and $C$ should be statistically relevant both for $A$ and $B$.

Reichenbach proceeds to point out two explanatory features of the proposed common causes. The first one is that from the conditions (1.3)-(1.6) the correlation between $A$ and $B$ is *deducible*. (We shall investigate this and related ideas in section 2.1.) This fact is interpreted by Reichenbach as meaning that the fork $ACB$ "makes the conjunction of the two events $A$ and $B$ more frequent than it would be for independent events" (*ibid.*), and that is why the author proposes to call such forks *conjunctive forks*. The second explanatory feature of common causes is that, due to screening off, the correlation in a sense "disappears"—"relative to the cause $C$ the events $A$ and $B$ are mutually independent" (*ibid.*). Due to these features, a common cause makes it possible to derive statistical dependence from an independence. The common cause is therefore the "connecting link", and the conjunctive fork "is therefore the statistical model of the relationship formulated in the principle of the common cause" (*ibid.*, p. 160).

What follows next is the proof of the above mentioned fact that from conditions (1.3)-(1.6) one can derive the correlation between $A$ and $B$. It is thus quite puzzling why, on the next page (163), Reichenbach writes "These results may be summarized in terms of the principle of the common cause (...)". Which results? So far, the existence of common causes as the middle links in conjunctive forks was distinctively assumed, not reached as any sort of result. What is more important now, though, since the author attempts a justification of the Principle later on, is its formulation (reworded so it would not refer to equations in Reichenbach's text by their numbers):

**Reichenbach's PCC—the "correlation" formulation:** "If coincidences of two events $A$ and $B$ occur more frequently than would correspond to their independent occurrence, that is, if the events are correlated, then there exists a common cause $C$ for these events such that the fork $ACB$ is conjunctive."

Notice that with the move from speaking about single coincidences to correlations the word "improbable" disappears. In the above formulation there is no division between "probable" and "improbable" (or "unexpected" / "accidental") correlations. Common causes are to exist for *all* correlations.

What is, for Reichenbach, the relationship between the statistical conditions (1.3)-(1.6) and the concept of common cause? Being a common cause of $A$ and $B$ is not sufficient for being the middle link of a conjunctive fork: $A$ and $B$ may simply not be correlated (Reichenbach's example is that of two dice being thrown by the same hand). In the other direction, being the middle element of a conjunctive fork is for Reichenbach certainly not sufficient for being a common cause, since common effects may also satisfy conditions (1.3)-(1.6). However, for his idea for defining time direction to work, it is absolutely crucial to ascertain that if a conjunctive fork $ACB$ is open[6], then $C$ is a common cause of $A$ and $B$ and not their effect. In this way he will be able to frame his definition of time direction in terms of macrostatistics as "In a conjunctive fork $ACB$ which is open on one side, $C$ is earlier than $A$ or $B$" (*ibid.*, p. 162). But does he succeed in showing the causal asymmetry of conjunctive forks? This may initially seem to be a side issue for the Principle of the common cause, but it is not: an example of a conjunctive fork open to one side, containing two events and their common effect, such that there is no common cause for the two events which together with them constitutes a conjunctive fork, would be a counterexample to the Principle. The fact that the issue was discussed in this context by perhaps the staunchest proponent of the principle, Wesley Salmon (1984), is another reason for which we will return to it in one of the coming sections.

### 1.3.1 Reichenbach's Argument for the Principle

What is, then, the justification given by Reichenbach for his Principle? It is supposed to follow from the second law of thermodynamics—the entropy of an isolated system which is not in equilibrium tends to increase—supplemented with an additional assumption, labeled "branch hypothesis", which we shall now consider.

As we said earlier, in his book Reichenbach does not use the formalism of probability spaces which has since then become the standard approach. Instead in §12 he introduces the so called "Probability Lattice". In the context of the book, it is a mathematical construction for describing processes of mixture. A probability lattice is a two-dimensional matrix; each row represents the history of a single object, e.g. a molecule of gas (thus it is also called a "time ensemble"), and each column is a time-slice through the system under consideration, containing information about the state of all molecules in the system at a given time (being thus also called a "space ensemble"). To use Reichenbach's own example, consider a container with two compartments—$L$ and $R$—and assume there are molecules of nitrogen in compartment $L$ and oxygen in compartment $R$. Suppose the wall dividing the compartments is removed and the substances begin to mix with each other. If we restrict our attention to nitrogen only, and record only the positions of the molecules (in a binary way, "$L$"

6 This seems to mean that one of two possibilities (one of which is almost immediately excluded) occurs: either (1) $C$ is a common cause of $A$ and $B$, and there exists no common effect $D$ of $A$ and $B$ such that $ADB$ would constitute a conjunctive fork, or (2) $C$ is a common effect of $A$ and $B$, and there exists no common cause $D$ of $A$ and $B$ such that $ADB$ would constitute a conjunctive fork.

or "$R$"), the first column of our probability lattice should be filled exclusively with $L$s, while the farther we go to the right, the more the proportion of $L$s and $R$s in a given column approaches 1/2.

The lattice will be a "lattice of mixture" (Reichenbach (1971), p. 103) only if it meets a few conditions, discussed on pages 100-103 of the book. The two simpler ones regard the initial column (which should be ordered[7]) and the aftereffect in rows (an $R$ at position $i$ in a row increases the chance for an $R$ at position $i+1$ in the same row). The other two, though, "independence of rows" and (especially) "lattice invariance" (which allows making inferences "from the time ensemble to the space ensemble"), are highly non-trivial. It would not be proper to study the conditions here in detail, since they are not at the heart of Reichenbach's argument for the PCC.[8]

The formalism was needed because the branch hypothesis itself refers to a lattice of mixture (Reichenbach (1971), p. 156). The general idea is that the universe, as a whole, is a system with currently low entropy but which increases over time (barring some short-term anomalies). From the main system, smaller systems branch off, and are isolated for a certain period—but they are connected with the main system at both ends. The entropy of these "branch systems" is also (in general) low at one of these points and high at the other; the crucial thing is that the direction towards higher entropy is in general parallel throughout the branch systems. This covers four out of five assumptions making up the hypothesis; the remaining one is that the lattice of branch systems is a lattice of mixture.

Suppose, for now, the branch hypothesis is true. How should the PCC follow? Reichenbach first tries to show (pp. 164-165) how a specific type of ensemble gives rise to a conjunctive fork, transcribed into his lattice notation. This ensemble containst two types of branch systems, $T_A$ and $T_B$ (systems of the first type may assume state $A$ and the others state $B$), such that sometimes a system of type $T_A$ and another of type $T_B$ coincide in their first state (call it $C$). A composite probability lattice is constructed from the two lattices for two system types by appropriately "gluing" some rows on top of others so that for any case of the above mentioned coincidence the row for the system of type $T_A$ is on top of the row for the system of type $T_B$. For future reference, let us

7 In the sense that it should illustrate a state of order; just like in the previous example, the initial—ordered—state of the system is illustrated by a column with the letter $L$ in all entries.

8 But it has to be noted that while there may be some intuitive appeal of those two conditions being connected with a mixing process, the author himself struggles with his own notation, being forced to use sub-subscripts, and we hope the Reader who consults the book will agree that it is not evident that Reichenbach's formulas in his lattice-lingo adequately express what he says in English. (For example, why does the right-hand side of formula (17) (p. 101) express any "vertical probability" (as defined on p. 99) at all?) Even if these difficulties were dispensed with, there is no justification for lattice invariance save for a reference to Reichenbach (1949), where it is stated that the kinetic theory of gases makes a similar assumption (p. 174).

state that his goal here was to ascertain that "whenever two causal lines leading to $A$ and $B$ are connected by their first element $C$, the fork $ACB$ is conjunctive" (*).

Then Reichenbach claims that "the branch hypothesis tells us that if a state occurs more frequently in the space ensemble than corresponds to a certain standard, namely, to its probability in the time ensemble, there must have existed an interaction state in the past" (p. 166) (**). This sentence is difficult to grasp due to its lack of quantifiers over columns and rows. Should we read it as "the branch hypothesis tells us that if in a lattice of mixture there exist row $k$ and column $i$ such that a certain state occurs more frequently in $k$ than in $i$, (...)" or "the branch hypothesis tells us that if in a given lattice of mixture it is true that for any row $k$ and column $i$ a certain state occurs more frequently in $k$ than in $i$, (...)"? The fact that in the previous paragraph we seem to have been actually considering three-dimensional probability lattices[9] does not help, either.

But let us, again, drop this issue (and the issue of whether the above actually follows from the branch hypothesis). The next step in Reichenbach's reasoning is that

> "If the two causal lines leading to $A$ and $B$ were not connected by their first elements, the probability of the joint occurrence would be given by $P(A) \cdot P(B)$."[10] (***)

This, unfortunately, begs the question. By contraposition and combination with (*) we get: "if $A$ and $B$ are correlated or anti-correlated, the two causal lines leading to $A$ and $B$ are connected by their first element $C$, and the fork $ACB$ is conjunctive ", which is at first sight an even stronger statement than the PCC.[11] Statement (***) needs to be backed up, but is not. It cannot be backed up by (**)—because, however we understand it, it is an implication from the fact that the probability of a given state in the space ensemble is different (higher) than its probability in the time ensemble, without reference to the actual values of the probabilities! So, *a priori*, it is consistent with (**) that for some $A$ and $B$, the causal lines leading to $A$ and $B$ are not connected by their first elements, but the probability of the joint occurrence is given by $P(A) \cdot P(B)+0.05$, which is inconsistent with (***).[12] Sadly, we have to conclude the argument

---

9 The additional dimension, apart from rows and columns, stems from the fact that rows from the lattice for systems of type $T_A$ are "above" the ones for systems of type $T_B$; it cannot be the two-dimensional sort of "above" used in statements like "on this very page, the previous line is above this one", since were it so it wouldn't be possible for $A$s and $B$s to happen in the same row of the composite lattice, which is explicitly required by Reichenbach's mathematical formulas.

10 *ibid.*

11 Only at first sight, because if events $A$ and $B$ are anti-correlated, then $A$ and $B^{\perp}$ are positively correlated (and *vice versa*), so the PCC may also be read as demanding explanation for anti-correlations.

12 The point will be perhaps more palatable if made colloquially: everyone remembers that "correlation does not mean causation". But (since $A$ and $B$, belonging by assumption to isolated branch systems, cannot cause one another) (***) says basically that "absence of causation means absence of correlation"! The author, when claiming (***), has to have in mind something similar to the negation

given by Reichenbach misses a link without which part (***) assumes the thesis. Thus the status of the Principle of the common cause in *The Direction of Time* is still that of a hypothesis. It does not change the fact that it may well be a valuable rule of human reasoning; simply, Reichenbach does not succeed in showing it to be a well-proved theorem.

It has to be added that Reichenbach himself thought that the PCC "reiterates the very principle which expresses the nucleus of the hypothesis of the branch structure" (*ibid.*, p. 167). Perhaps, then, no separate argument for the PCC is needed and a justification of the hypothesis of the branch structure would suffice. We will argue in the next section that this prospect is sadly also not hopeful, among a few other drawbacks of Reichenbach's account.

### 1.3.2 Other Problems with Reichenbach's Approach

#### 1.3.2.1 The Hypothesis of the Branch Structure: "Main System" and Entropy

The first worry regarding the hypothesis concerns what it is that is supposed to branch. Reichenbach offers a few illustrations. In the first one (fig. 20 from Reichenbach (1971), p. 119) we are supposed to see a "long upgrade of the entropy curve of the universe" and "systems branching off from this upgrade, assuming that these branch systems remain isolated for an infinite time" (*ibid.*, p. 118). The second one (fig. 21 from Reichenbach (1971), p. 127)) differs in that the systems which branch off from the main system return to it and that it contains also a downgrade of the entropy curve.

In both images the vertical axis is supposed to depict entropy. And the problem is that, while not all concepts of entropy are that of an additive quality (see e.g. Palm (1978)), the types of entropy considered by Reichenbach are additive, as he says himself on p. 53 ("If two systems are brought together, their entropies are additive"). Therefore, suppose the universe consisted of a system which from time to time divides into two systems that remain isolated for a certain period and then connect again. It seems to me that Reichenbach's fig. 21 is incompatible with additivity of entropy—when the two systems reconnect, the entropy of the whole is depicted as being equal to the entropy of any of the two, while it should be the sum of the two. A picture which would do justice to the additivity of entropy would not contain a branching curve, though.

Perhaps this is just a pictorial difficulty of no greater import. But the bigger problem with the hypothesis is that it refers to "the main system"; presumably, "the main system of the universe". It is never made clear what the main system is. Is Earth a part

---

of the hackneyed slogan; namely, that correlation *does* mean causation, if not between the correlated events (since they occur at the same time or belong to isolated systems), but between them and their common cause. This is yet another informal statement of Reichenbach's PCC.

of it, or is the humanity in some backwater region of the universe? At first sight, the concept of the main system is important for the hypothesis; the main system is to serve as the root from which the other systems branch, and to which they eventually return. On the other hand, perhaps the hypothesis could be reformulated so that it would refer to an ensemble of systems whose both ending points are in other systems, and which are isolated from all other systems apart from their endpoints. In this case, there would be no distinguished "root", or "main", system—and similarly, there would be no need to use the name "branch system" instead of simply saying "system": all systems would have equal rights, so to speak. One would also have to take care when accommodating the old Assumption 4 ("In the vast majority of branch systems, one end is a low point, the other a high point") to the new hypothesis; what if a system $K$ branches off a system $L$ at a point of $L$'s high entropy, but, after a period of isolation, connects with a system $M$ at a point of $M$'s low entropy? I do not think these difficulties are insurmountable. It is feasible that one could reformulate Reichenbach's hypothesis of the branch structure so that it would not refer to any "main system", while still capturing as much of the intentions of the original author as possible. Then the task of deriving the PCC could be approached again. A problem with this is that one would still be trapped with Reichenbach's probability lattice approach and his notation. We prefer to pursue another option and consider the chances of proving theorems related to the PCC using the machinery of "modern" probability theory. This endeavour is taken up in chapter 5.

#### 1.3.2.2 What do the Initial Examples Illustrate?

As we said earlier on, Reichenbach himself claims that in his book probability is always to be understood as a limit of a relative frequency. This would seem to preclude ascribing probability to "token", unrepeatable events; in other words, there should be no "single-case probabilities". However, we already quoted passages from Reichenbach (1949) indicating that there is an, albeit elliptic, way of speaking about constructs which are to serve as a substitute for them in Reichenbach's theory. How should we, then, understand the initial examples of common-causal reasoning offered by the author (and quoted here on p. 11), the fire and the wind, the burned-out bulbs, and the ill actors? The common cause is invoked after the occurrence of a *single* event is observed. At the end of section 1.1.2 we claimed that no beliefs about the probability of such an event should be formed just because of a single occurrence. Reichenbach seems to agree, writing on p. 158, not long after the examples have been presented, "(...) we assume that $A$ and $B$ have been observed frequently; *thus it is possible* to speak about probabilities $P(A)$, $P(B)$ and $P(B \mid A)$ (...)". So, in the initial examples we are not supposed to think of probabilities, let alone correlations. Therefore they cannot be of any support for the Principle in its "correlation" formulation; they only illustrate the "coincidence" formulation in action.

Remember, though, that the two features Reichenbach advertised as due to which a common cause has explanatory value stem from the common cause being a middle link in a conjunctive fork. Since the definition of the fork is probabilistic, if we know nothing about the probability of the given common cause $C$, we cannot judge whether it is the middle link in a conjunctive fork $ACB$, and so cannot benefit from the above-mentioned features: (1) that the correlation disappears when the events $A$ and $B$ are considered conditional on $C$, and (2) that the correlation is derivable from the conjunctive fork condition. These two features show us why the PCC may be promoted as one of the principles guiding the human search for explanation, but *only in its formulation referring to a "correlation"* (p. 13), not in the one bringing up an "improbable coincidence" (p. 11).

In conclusion, Reichenbach's initial examples illustrate only the "coincidence" formulation of the Principle, which lacks the important explanatory features of the "correlation" formulation.

#### 1.3.2.3 On Forks Open to the Past

First let us ask about sufficient conditions for a triple of events $ACB$ to constitute a conjunctive fork. Are the statistical requirements (1.3)-(1.6) enough? Consider some events $A$, $B$ and their common cause $C$, which operates in a deterministic way: $P(A \mid C) = P(B \mid C) = 1$, $P(A \mid C^{\perp}) = P(B \mid C^{\perp}) = 0$. Notice that

$$\begin{aligned} P(AC \mid B) &= 1 = P(A \mid B)P(C \mid B); \\ P(AC \mid B^{\perp}) &= 0 = P(A \mid B^{\perp})P(C \mid B^{\perp}); \\ P(A \mid B) &= 1 > 0 = P(A \mid B^{\perp}); \\ P(C \mid B) &= 1 > 0 = P(C \mid B^{\perp}), \end{aligned}$$

so the triple $ABC$ satisfies the statistical requirements for being a conjunctive fork, with $B$ being the middle link. But, if forks are to represent causal relations, then $ABC$ cannot be a conjunctive fork, because it is not a fork in the first place. The moral is this: prior causal knowledge is needed to determine whether the fact that a triple of events satisfies conditions (1.3)-(1.6) means that the triple constitutes a conjunctive fork. We differ in this opinion from e.g. Salmon, who considers the statistical conditions as definitional[13] for the notion of the conjunctive fork, but assumes (for unrelated reasons) that none of the probabilities occurring in the requirements may be equal to 0 or 1. His reasoning would thus be unaffected by my counterexample.

Conjunctive forks open to the past would of course (just as any examples of two correlated events with neither a common effect nor a common cause) constitute coun-

13 See Salmon (1984), p. 159-160.

terexamples to the PCC. Reichenbach claims that whenever a conjunctive fork $AEB$ is found such that $E$ is a common effect of $A$ and $B$, there exists an event $D$, which is a common cause of $A$ and $B$, and is the middle link of a conjunctive fork $ADB$. Thus there exist no conjunctive forks open to the past.

Reichenbach offers both a general argument and some specific examples. The argument is of a teleological nature and refers to the fact that we do not accept final causes as explanations. Final causes are deemed incompatible with the second law of thermodynamics in the preceding chapter (§18 of Reichenbach (1971)); a general question is asked: how are we to explain the presence of a highly ordered (and so, very improbable) state of a system (such as a trace of footprints in the sand)? Reichenbach's answer is that we are supposed to look for an interaction "at the lower end of the branch run through by an isolated system which displays order", which will be the cause; "the state of order is the effect" (p. 151). The ordered state is, then, to be understood as a *post-interactive* state. Since the overarching goal is to provide a definition of time direction (as we have seen Reichenbach doing in the following chapter—§19 of Reichenbach (1971)—by defining what is to be meant by "past"), the author proposes to consider the system containing the beach with the footprints in "reverse time" (p. 153). We would have to think of the ordered state as a *pre-interactive* state, and so, in our search for its explanation would end up with a final cause ("The wind transforms the molds in the sand into the shapes of human feet *in order that* they will fit the man's feet when he comes", *ibid.*); in general, we would "explain the improbable coincidences by their *purpose* rather than by their *cause*" (*ibid.*). Since this is implausible, the conclusion is that the direction of time should be defined, generally speaking, from interaction to order, rather than the other way round. And so, "if we define the direction of time in the usual sense, there is no finality, and only causality is accepted as constituting explanation" (p. 154).

Unfortunately, the nonexistence of conjunctive forks open to the past would follow from the above only had it been established that such a conjunctive fork would necessitate the usage of final causes. This would only be the case if (1) every correlation between events having a common effect but no common cause (i.e. events being the extreme elements of a causal fork open to the past) had an explanation; (2) the only accepted way of explaining such a correlation would be to refer to an event in their causal future. But Reichenbach does not give arguments for any statements similar to the two above; in fact, he seems to rely on an (unsupported) fundamental principle that every correlation whatsoever has an explanation. Notice also the curious jump from the epistemic to ontological perspective on p. 163: "A common effect cannot be regarded as an explanation and thus need not exist". In general, it does not seem that Reichenbach's general argument for the nonexistence of open conjunctive forks with a common effect as the middle element holds up under scrutiny, mainly due to the trick of deriving the ontological conclusion from epistemic premises (like the universal requirement for explanation for correlations).

Coming now to the specific examples, the author gives an instance of a fork open to the past on p. 163, aiming to convince the Reader that the fork cannot be conjunctive. Let us quote a part of the example:

> "For instance, when two trucks going in opposite directions along the highway approach each other, their drivers usually exchange greetings, sometimes by turning their headlights on and off. We have here a fork $AEB$, where $E$ is the exchange of greetings, which is a common effect of the "coincidence" of the trucks, that is, of the events $A$ and $B$" (Reichenbach (1971), p. 163).

It is not evident how we should think about probabilities in this case, but one way would be to hold fixed a fragment $X$ of some highway, and let $A$ be the event "there is a truck going in the eastern direction in the fragment $X$", $B$ be the event "there is a truck going in the western direction in the fragment $X$", and $E$ "two trucks going in the opposite directions in the fragment $X$ are flashing their headlights". We can check whether the events occur e.g. every second. Then it is very likely that $E^{\perp}$ does not screen off $A$ from $B$, so the three events indeed do not form a conjunctive fork. Still, a general argument against the mere possibility of such a fork open to the past is needed.

A related problem appears in Salmon (1984), where on p. 164-165 an example offered by Frank Jackson of a conjunctive fork open to the past is discussed.

> "[C]onsider a case that involves Hansen's disease (leprosy). One of the traditional ways of dealing with this illness was by segregating its victims in colonies. Suppose that Adams has Hansen's disease ($A$) and Baker also has it ($B$). Previous to contracting the disease, Adams and Baker had never lived in proximity to one another, and there is no victim of the disease with whom both had been in contact. We may therefore assume that there is no common cause. Subsequently, however, Adams and Baker are transported to a colony, where both are treated with chaulmoogra oil (the traditional treatment). The fact that both Adams and Baker are in the colony and exposed to chaulmoogra oil is a common effect of the fact that each of them has Hansen's disease. This situation, according to Jackson, constitutes a conjunctive fork $A$, $E$, $B$, where we have a common effect $E$, but no common cause" (Salmon (1984), p. 164)

To check whether the statistical conditions are satisfied, one has of course to check e.g. probabilities $P(A \mid E^{\perp})$ and $P(A^{\perp} \mid E^{\perp})$. But how should we do this? We had already assumed that Adams has Hansen's disease and that he is in the colony. How can we ask about the probability that he is *not* ill or that he is *not* in the colony? Certainly we are not evaluating a probability of a counterfactual statement[14]. Instead, it is evident from p. 165 of Salmon (1984) that the author calculates the probability $P(B^{\perp} \mid E)$ simply by taking the proportion of people in the colony who are not ill (the medical personnel) to all members of the colony. But in this way he transforms a constant into

14 Which is a task attempted later on e.g. in chapter 7 of Pearl (2000).

a variable and it is no longer possible to differentiate between events $A$ and $B$, since both of them are "a randomly chosen man from the colony has Hansen's disease".

It would seem, then, that Reichenbach's account lacks a general argument for his point, and Salmon's considerations on the subject are defective. On the other hand, we have to admit we have been unable to find a "real-world" example of a conjunctive fork open to the (causal) past. Still, consider the following hypothetical situation: a group of 10000 men (labeled, for our convenience, from 1 to 10000) considered as representative for the region is tested for hypocalcemia ($E$), lactose intolerance ($A$) and hypoparathyroidism ($B$). Lactose intolerance and hypoparathyroidism have no known common cause, while it is known that each may lead to hypocalcemia. If:

- men labeled from 1 to 1000 (and only them) have hypocalcemia;
- men labeled from 1 to 500 and from 1001 to 4000 (and only them) have lactose intolerance;
- men labeled from 251 to 750 and from 3001 to 6000 (and only them) have hypoparathyroidism;

then it is straightforward to see (if we accept the move from relative frequencies to probability: here, for the sake of the example, we can simply say that the population from which the sample had been drawn is identical to the sample) that the fork $AEB$ satisfies the requirements from the definition of a conjunctive fork (1.3)-(1.6). However, the middle element of the fork is a common effect of the two other elements, which have no known common cause. I do not see why such situations should be impossible; yet again, I have been unable to find a "real" example.[15]

(A different matter is whether a "conjunctive fork open to the past" and a "conjunctive fork $AEB$ with the middle element $E$ being a common effect of $A$ and $B$, such that there is no common cause $C$ creating a conjunctive fork $ACB$" are to be identified. They certainly are on the assumption that "past" in the first expression is to be understood as "causal past").

## 1.4 The PCC After Reichenbach

Reichenbach's Principle was heavily promoted in the 70s and 80s by Wesley Salmon (e.g. Salmon (1971)). More recently, it has been an inspiration for a fundamental condition in the field of representing causal relations by means of directed acyclic graphs (see chapter 4). However, a plethora of counterarguments appeared; most are gathered and discussed in Arntzenius (1992). Some, e.g. Sober's (1988) "sea levels vs. bread prices" argument, were directed against any sort of general requirement of

15 Another hypothetical example of a conjunctive fork not pointing to a common cause was also presented in Torretti (1987).

common causal explanation. Others led a few philosophers (e.g. Salmon (1998b)[16] and Cartwright (1988)) to transform Reichenbach's idea, preserving the Principle's requirement of common causes for correlations, but changing the screening off condition, or supplementing it with other conditions. It would be of no use for the current essay to discuss all these ideas in detail: our focus is on the notions of common cause revolving around the original idea of screening off. We will however describe the three arguments we would rate as most important. These are:
- the argument from Bell inequalities, to which we will devote the whole chapter 3;
- the argument from conservation principles, described in section 1.4.1;
- and the "sea levels vs. bread prices" argument, described in section 1.4.2.

Later, starting in the 90s, Reichenbach's idea in the form of PCC 4 was defended in papers by M. Rédei, G. Hofer-Szabó and L. Szabó (e.g. Hofer-Szabó et al. (2000)): rather than confronting the earlier counterarguments to Reichenbach's idea directly, the authors proposed mathematical arguments in favour of PCC 4. It is to this area of research that the current study aims to contribute in chapters 5 and 6. Let us first describe the two arguments against Reichenbach's Principle we just mentioned above.

### 1.4.1 The Argument from Conservation Principles

We will cite the formulation of this argument given in Arntzenius (1992), since it seems to be the most concise:[17]

> "Suppose that a particle decays into 2 parts, that conservation of total momentum obtains, and that it is not determined by the prior state of the particle what the momentum of each part will be after the decay. By conservation, the momentum of one part will be determined by the momentum of the other part. By indeterminism, the prior state of the particle will not determine what the momenta of each part will be after the decay. Thus there is no prior screener off." (Arntzenius (1992), p. 227-8.)

There are numerous variants of this argument in the literature; the version from Salmon (1998b) refers to Compton scattering. In the same paper Salmon, as an answer to the problem, proposes the introduction of another kind of fork (apart from the conjunctive variety), the so called *interactive* fork. Probabilistically, an interactive fork with the middle element $C$ and two extreme elements $A$ and $B$ differs from a corresponding conjunctive fork in that instead of the two screening off requirements a single condition is introduced: namely, $P(AB|C) > P(A|C)P(B|C)$. That it is met by

16 Originally published in 1978.

17 It is labeled "Indeterministic Decay with Conservation of Momentum" and attributed to van Fraassen (1980).

the examples built around some conservation principle becomes evident when we notice that in such examples (if $C$ is the state of the compound before the splitting) $1 = P(A|B \wedge C) > P(A|C)$.

Notice that the argument only implicitly refers to probability, *via* the notion of screener-off. No probability spaces are defined. Therefore it is an argument against PCC 2. It is not clear what force it would have against PCC 4, which—as mentioned above—has been mathematically proven to be true. One would have to consider all probability spaces which could be used to describe the decay event and its consequences; then the extensions of those spaces which contain common causes in the sense of PCC 3; and finally ponder the question whether such events could have anything to do with what we would naturally accept as a common cause of the properties of the two particles.

An *ad hoc* solution—but not without intuitive merit—on part of a proponent of PCC 2 could be that correlations which arise due to conservation principles do not demand (additional) explanation: if we know the principle at work, we do not require anything more to explain the correlation. Salmon's way out (philosophically rooted in the distinction between causal processes and interactions) was simply to incorporate interactive forks into the picture and to say that some correlations are explained by events which together with the correlated events form a conjunctive fork, but some others demand as their *explanantes* the middle elements of interactive forks.[18]

### 1.4.2 The "Sea Levels vs. Bread Prices" Argument

This argument first appeared in Sober (1988) and was elaborated in Sober (2001). The most important thing is that, in the parlance of the current essay, it is an argument for abandoning PCC 1 for PCC 2: not all correlations demand a common causal explanation.[19] By *reductio*: otherwise a correlation between Venetian sea levels and British bread prices would demand such an explanation, while it surely does not. More extensively:

> "Consider the fact that the sea level in Venice and the cost of bread in Britain have both been on the rise in the past two centuries. Both, let us suppose, have monotonically increased. Imagine that we put this data in the form of a chronological list; for each date, we list the Venetian sea level and the going price of British bread. Because both quantities have increased steadily with time, it is true that higher than average sea levels tend to be associated with higher than average bread prices. The two quantities are very strongly positively correlated.

18 In fact, Salmon himself eventually espoused a variant of the "conserved quantity" theory, see Salmon (1998a).

19 Which demand and which do not is in Sober's account decided by our background theory.

I take it that we do not feel driven to explain this correlation by postulating a common cause. Rather, we regard Venetian sea levels and British bread prices as both increasing for somewhat isolated endogenous reasons." (Sober (1988), p. 215.)

There are several strands of thought in the literature on the argument. We will try to label and shortly discuss them.

1. **No correlation at the level of changes.** Forster (1988) was the first to notice that, while the sea levels and bread prices are correlated, their respective changes are not: year by year, both the former and the latter increase. It is hard to estimate the import of this observation: in fact, the lack of correlation on the level of changes would perhaps intuitively indicate lack of causal connection, which would strengthen Sober's point. In any case, Sober (2001) presented an example stemming from evolutionary biology in which the correlation persists on the level of changes of the values of two attributes, and in which also no common causal explanation is expected.
2. **Mixing.** Some authors (e.g. Spirtes et al. (2000), p. 33-37) point out—going back to the work of G.U. Yule in the beginning of the XXth century—that a correlation between attributes in a population may be the result of mixing two populations in which the attributes are not correlated. The solution of Spirtes, Glymour and Scheines, working in the formalism of directed acyclic causal graphs (more on that in chapter 4), is to treat "belonging to one of the given subpopulations" as an attribute in the "big" population, which enables them to recover the proper conditional independencies. This amounts to saying that whenever such a mixing occurs, the correlation is explained by the mixing itself. The authors treat (p. 37) Sober's example as a case of mixing (in fact, they claim that his point was similar to Yule's) and consider the case closed. This is consistent with PCC 2: if we know that such a mixing occurred, then we either do not think of the correlation as demanding an explanation at all, or we consider it explained by the mixing itself. However, the particular example of sea levels and bread prices does not lend itself easily to the "mixing" interpretation. What are the two populations to be mixed? They cannot be the 200 years, since there is only one set of years to be considered. Thinking of Sober's example as a case of mixing seems to require some serious mind-twisting. It might be better to look at it as consisting of two monotonically increasing time series.
3. **Time series.** Hoover (2003) observes that the data given by Sober allow us to infer that there is a correlation on the level of frequencies, but it does not necessarily follow from that that there is a correlation on the level of probabilities. "(...) most statistical inference and most of our own probabilistic intuitions are based on stationary probability distributions" (p. 532).
   A rigorous discussion of this point would require numerous definitions, so let us settle for a more informal account. A necessary condition for a time series to be *stationary* is that "the covariance between the values of the series at different

> times depends only on the temporal distance between them" (p. 532). Of course, a monotonically increasing time series is not stationary. However, for any time series we can consider a series of *differences* between the consecutive values of the series. It may very well happen that the series is stationary (consider e.g. the case of the series of consecutive natural numbers from 1 to 200); in that case the original series is said to be *integrated of order* 1, or "I(1)". As Hoover points out (p. 545-546), a linear combination of two I(1) time series is in general also I(1), but it may happen that there exists a linear combination which is stationary: only in this case (in which we say the time series are *cointegrated*) the data constitute evidence for probabilistic dependence. As Reiss (2007) puts it (p. 184), "inferring from a sample correlation to a probabilistic dependence means that one takes the most likely data-generating process to be stationary". In Sober's case we are likely to assume otherwise. Hoover proposes to let the PCC apply to cases in which the correlated series are either (1) both stationary or (2) both I(1), but cointegrated. Since Sober's time series are not cointegrated, they do not constitute a counterexample to Hoover's version of the principle.

The general direction of these replies against the "sea levels vs. bread prices" argument seems to be that, while Sober's paper could be interpreted as containing a probability space with unexplained correlations, it is doubtful whether that space corresponds to anything "in the world"; whether there *really* are factors correlated in the way the probability space would have it. Inspection of a mathematical object, even a probability space constructed on the basis of actual, observed frequencies, is not enough to disprove the Common Cause Principle in its PCC 1 formulation.

We would like to divert attention to a different issue. Why do we think the correlation between Venetian sea levels and British bread prices does not demand explanation? Is there more to say on the topic than Hoover's idea of referring to the two time series being nonstationary but not cointegrated?

### 1.4.3 Which Correlations Demand Explanation?

Various authors have fleshed out the beginning part of PCC 2 differently. Some say that only "improbable" correlations demand explanation. This can mean simply "statistically significant" (Forster (1988), p. 539), "unexpected or surprising" (Uffink (1999)), or "such that the assumption that it arises from causally unrelated processes will render it unexpected", with a formally defined notion of "unexpected" (Berkovitz (2000), p. 65).

There is a sense, meta-probabilistic in a way, in which a correlation can be rigorously thought of as "improbable". Consider a finite probability space $\langle \Omega, \mathcal{F}, P \rangle$, in which $\Omega$ has $n$ elements, $\mathcal{F}$ is the set of all subsets of $\Omega$, and for any $A \in \mathcal{F}$, $P(A) = \frac{card(A)}{n}$. Suppose two subsets $A$ and $B$ of $\Omega$ are chosen at random in the sense that every member of $\Omega$ has a chance of $\frac{1}{2}$ of belonging to $A$ and the same chance of belonging to $B$. It can be checked using a moderate dose of combinatorics and Stirling's approximation that with $n$ approaching infinity, the chance of arriving at a probabilistically independent pair by the process just outlined approaches 0. Informally, we would say that "in a big enough population, (almost) everything is correlated with (almost) everything". But the proportion of pairs such that $|P(AB) - P(A)P(B)| < 0.05$ (which is one, quite arbitrary, way of saying "weakly correlated pairs") to all pairs increases with $n$, too—strongly correlated pairs are infrequent in this sense (i.e., it is hard to come upon them by pure chance). (I have to confess that for this statement I only have an argument of "consulting statistical software"—having used the "R" software[20] to track the proportions of weakly and strongly correlated pairs in populations of increasing sizes.) So, if we do not possess any knowledge about the genesis of some two events, it should be natural to expect them to be correlated, but only weakly. The subjective degree of this expectation should vary with the size of the population involved. However, this approach is too abstract to properly illustrate our beliefs regarding the probabilistic nature of phenomena we observe in the world: we in general do not start with a "clean slate", but possess some background knowledge which influence our beliefs regarding such issues.

In his account[21], to the condition of "improbability" Berkovitz also adds, without further commentary, that the correlation should be "non-accidental" (p. 56). What does it mean for a correlation to be accidental? Perhaps it will be illuminating to consult a clarification of types of correlation from Haig (2003), where a reinterpretation of the notion of spuriousness is argued for.

A traditional (see e.g. Hitchcock (2010)) example of a spurious correlation is that of a barometer and a storm. Shortly speaking, a spurious correlation arises between events which are not directly causally connected, but have a common cause. Haig notices that the word "spurious" might be misleading in this case, since the correlation is due to a genuine causal connection—in contrast with the correlations which arise e.g. from a sampling bias. Haig's classification is presented in figure 1.2.

A "nonsense correlation" is that for which "no sensible, natural causal interpretation can be provided" (Haig (2003), p. 127). Haig's examples are "the high positive correlation between birth rate and number of storks for a period in Britain" and "the negative correlation between birth rate and road fatalities in Europe over a number of years". It is clear, I think, that Haig would interpret Sober's examples as falling into this category. "Spurious correlation" are these which are "not brought about by their claimed natural causes", but "by accident", for example due to "sample selection bias, use of an inappropriate correlation coefficient, large sample size, or errors of sampling, measurement, and computation" (Haig (2003), p. 128). Of the genuine cor-

**20** R Core Team (2012).
**21** Berkovitz (2000).

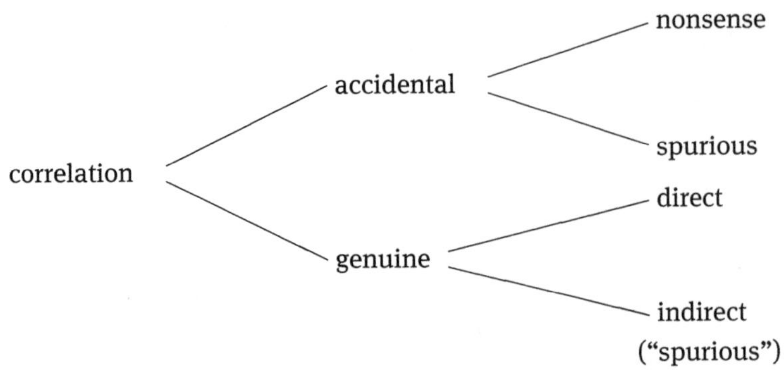


**Figure 1.2:** A classification of correlations from Haig (2003).

relations, the "indirect" are due to either "common or intervening causes" (p. 129)[22]. Notice that if we formulate our Principle as "any non-accidental correlation between events which are not directly causally related or logically dependent is due to a common cause" it is clearly true under Haig's account. Whether the common cause in question should screen off the correlated events is a different matter.

On the other hand, Sober would most likely say that—in the cases like the "sea levels vs. bread prices"—there is a perfectly natural causal interpretation which consists of two separate causal explanations of the two phenomena, and thus such correlations should not be counted as "nonsense". But let us examine this line of reasoning. Suppose that there is an explanation $E$ for the ongoing rise of sea levels; perhaps the melting of sub-polar glaciers. $E$ explains why the data for sea levels in Venice form a monotonic time series. Likewise, suppose there is some explanation $F$ for the ongoing rise of bread prices; perhaps a combination of high taxes and deteriorating crop levels. $F$ explains why the data for bread prices in Britain form a monotonic time series. But this does not yet explain the correlation in question. But why is $A$ higher than average in the given time period precisely whenever $B$ is? The only reason we can find is that the processes $E$ and $F$ are active in the same time period. If *this* is accidental, then the correlation between $A$ and $B$ should be deemed as accidental, too. Van Fraassen (1991, p. 350) uses the word "coincidence" in this context.

22 A correlation between $X$ and $Y$ can arise partially due to an intervening cause if there is a cause $Z$ which with $X$ jointly produces $Y$.

## 1.5 An Epistemic Position

It is interesting that in recent years two diametrically opposed approaches to the PCC have emerged: one (by Rédei, Hofer-Szabó and Szabó, discussed in chapter 6) we could label as "maximalist" aiming to prove that any correlation has an explanation by means of a common cause if the probability space is suitably chosen, and one perhaps aptly labeled as "minimalist", due to Reiss (2007). The latter renders the PCC as an intrinsically epistemic principle. Since it seems to have an illustrative purpose and amounts to saying "if there is no conflicting knowledge, correlation is evidence for causation", we will only give the formulation here without extensive discussion:

**PCC(Reiss).** "The proposition $e$ = "Random variables $X$ and $Y$ are (sample or empirically) correlated" is *prima facie* evidence for the hypothesis $h$ = "$X$ and $Y$ are causally connected". If all alternative hypotheses $h_i^a$ (e.g. "the correlation is due to sampling error", "the correlation is due to the data-generating processes for $X$ and $Y$ being non-stationary, "$X$ and $Y$ are logically, conceptually or mathematically related") can be ruled out, then $e$ is genuine evidence for $h$" (Reiss (2007), p. 193).

The epistemic nature of the Principle is evident from its use of the notion of "evidence"—no existential statements about the world are to be inferred from other statements of a similar nature; the Principle concerns the shaping of our conception of the causal structure of the world. The Principle is eminently fallible because we may simply have wrong evidence against the hypotheses alternative to the one of the existence of a common cause.

Before we tackle the issue of Bell inequalities, it is fitting to devote a chapter to the formal features of various notions related to screening off.

# 2 Screening Off and Explanation: Formal Properties

Arguably, the most important feature of Reichenbach-style common causes is that they screen off their effects. In this chapter we will consider a few constructions based on this idea which extend, or transform, Reichenbach's original notion.

A part of the literature refers to events defined as meeting Reichenbach's conjunctive fork conditions as ("Reichenbachian") "common causes"; see e.g. Hofer-Szabó and Rédei (2004). Hofer-Szabó et al. (2000) and Hofer-Szabó and Rédei (2006) even go so far as to state that Reichenbach himself defined common causes as the middle elements of conjunctive forks with correlated extreme elements; in other words, that fulfilling the statistical requirements for being the middle element of a conjunctive fork is sufficient to be a common cause for the correlated events. Due to the reasons already presented in section 1.3 we are reluctant to adhere to this tradition. Nevertheless, the main results of this work (see chapter 5) pertain to problems posed in various papers by the above-cited authors. Therefore, some slight terminological changes are in order.

**Definition 9 [Statistical Common Cause]** Let $\langle \Omega, \mathcal{F}, P \rangle$ be a probability space. Let $A, B \in \mathcal{F}$. If there exists $C \in \mathcal{F}$ different from both $A$ and $B$ such that

$$\begin{aligned}
P(AB \mid C) &= P(A \mid C)P(B \mid C);\\
P(AB \mid C^{\perp}) &= P(A \mid C^{\perp})P(B \mid C^{\perp});\\
P(A \mid C) &> P(A \mid C^{\perp});\\
P(B \mid C) &> P(B \mid C^{\perp}),
\end{aligned}$$

then $C$ is called a *statistical common cause* of $A$ and $B$.

It is intuitive that a similar notion could be considered, with the difference that it would permit the cause to be more complicated than a simple "yes" / "no" event. This is indeed the path taken without further comment by van Fraassen (1982), but only the screening off requirement is retained. A generalization which takes into account also the conditions of statistical relevance was developed by Hofer-Szabó and Rédei (2004); the resulting constructs were called "Reichenbachian common cause systems", but, for reasons given above, we will abstain from the adjective "Reichenbachian".

In view of the definition of statistical relevance (definition 7, p. 5) we can say that a statistical common cause $C$ is positively statistically relevant for both its effects, or that in such a case the pair $\{C, C^{\perp}\}$ is statistically relevant for the same events; it would be wrong to say that $\{C, C^{\perp}\}$ is *positively* relevant, since one of its elements lowers the probability of the effects.

**Definition 10 [Statistical Common Cause System]** Let $\langle \Omega, \mathcal{F}, P \rangle$ be a probability space. A partition of unity of $\mathcal{F}$ is said to be a *statistical common cause system* (SCCS) for $A$ and $B$ if it satisfies the statistical relevance condition w.r.t. $A$ and $B$, all its members are different from both $A$ and $B$, and all its members are screener-offs for the pair.

The cardinality of the partition is called the *size* of the statistical common cause system.

Statistical common cause systems come in different cardinalities; it was proven in Hofer-Szabó and Rédei (2006) that SCCSs of arbitrary finite size exist. However, for some time it has not been clear which range of cardinalities is admissible. The problem whether infinite SCCSs exist is posed in Hofer-Szabó and Rédei (2004). We will now show that

**Theorem 1** (Wroński and Marczyk (2010)). *The greatest possible cardinality of an SCCS is* $\aleph_0$.

*Proof.* **1.** No uncountable SCCSs exist. To see this, suppose that in some probability space $\langle \Omega, \mathcal{F}, P \rangle$ an SCCS $\{C_i\}_{i \in I}$ of size greater than $\aleph_0$ exists. Since all elements of the SCCS are screener-offs, they have to have positive probability—otherwise the required conditional probabilities are not defined. But we know from measure theory (see e.g. Theorem 10.2 in Billingsley (1995), p. 162) that this is impossible, since the SCCS is a partition of unity of $\mathcal{F}$ and so the set of its non-zero probability members can only be countable.[1]

**2.** We will now construct an example of a countably infinite SCCS.

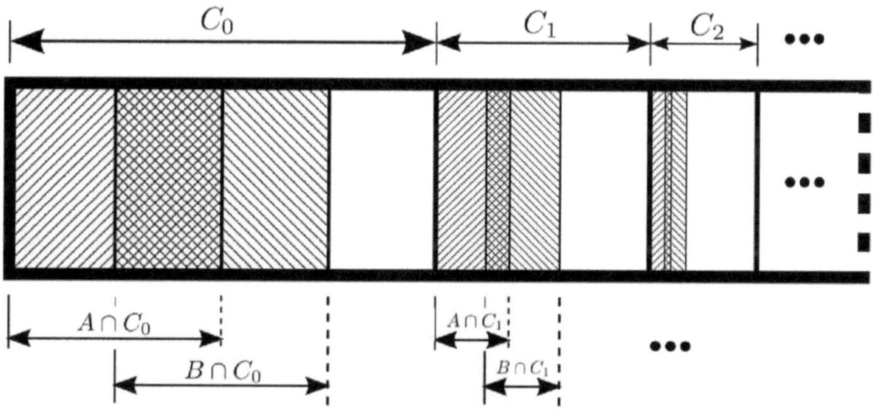


Let $\langle [0, 1), W, \lambda \rangle$ be a classical probability space with $W$ being the set of all Lebesgue-measurable subsets of the real interval $[0, 1)$ and $\lambda$ being the Lebesgue measure. Put

1 The Reader may prefer conditional probabilities given probability zero events to be always equal to 0, or 1 (see e.g. Adams (1998), p. 57). In these cases we note that for some distinct $k, l \in I$, $P(A \mid C_k) = P(A \mid C_l)$, which violates the statistical relevance condition.

$$C_n := \left[\frac{2^n - 1}{2^n}, \frac{2^{n+1} - 1}{2^{n+1}}\right);$$

$$C := \{C_n\}_{n \in \mathbb{N}}$$

It is evident that if $n \neq m$ $(n, m \in \mathbb{N})$, $C_n \cap C_m = \emptyset$ and that $\bigcup C = [0, 1)$, so $C$ is a countably infinite partition of $[0, 1)$. Notice that for any natural $n$, $\lambda(C_n) = \frac{1}{2^{n+1}}$.

For any $n \in \mathbb{N}$, we want both $\lambda(A \cap C_n)$ and $\lambda(B \cap C_n)$ to be equal to $\frac{1}{(n+2)\cdot 2^{n+1}}$. To improve the clarity of the notation below, put $l_n = \frac{1}{(n+2)\cdot 2^{n+1}}$. Define

$$A := \bigcup_{n=0}^{\infty} \left[\frac{2^n - 1}{2^n}, \frac{2^n - 1}{2^n} + l_n\right);$$

$$B := \bigcup_{n=0}^{\infty} \left[\frac{2^n - 1}{2^n} + \frac{n+1}{n+2} \cdot l_n, \quad \frac{2^n - 1}{2^n} + \frac{n+1}{n+2} \cdot l_n + l_n\right).$$

From the above definitions it follows that

$$\lambda(B \mid C_n) = \lambda(A \mid C_n) = \frac{\lambda(A \cap C_n)}{\lambda(C_n)} = \frac{\frac{1}{(n+2)\cdot 2^{n+1}}}{\frac{1}{2^{n+1}}} = \frac{1}{n+2};$$

whereas

$$\lambda(A \cap B \mid C_n) = \frac{\lambda(A \cap B \cap C_n)}{\lambda(C_n)} = \frac{\left(1 - \frac{n+1}{n+2}\right) \cdot \frac{1}{(n+2)\cdot 2^{n+1}}}{\frac{1}{2^{n+1}}} =$$

$$= \frac{\frac{1}{(n+2)^2 \cdot 2^{n+1}}}{\frac{1}{2^{n+1}}} = \frac{1}{(n+2)^2}$$

and so

$$\lambda(A \cap B \mid C_n) = \lambda(A \mid C_n)\lambda(B \mid C_n),$$

which means that $C$ satisfies the screening off condition.

Now, let $m, n \in \mathbb{N}$, $m \neq n$. Without loss of generality assume $m > n$. It follows that

$$\lambda(A \mid C_n) = \frac{1}{n+2} > \frac{1}{m+2} = \lambda(A \mid C_m)$$

and

$$\lambda(B \mid C_n) = \frac{1}{n+2} > \frac{1}{m+2} = \lambda(B \mid C_m).$$

Therefore the differences $\lambda(A \mid C_m) - \lambda(A \mid C_n)$ and $\lambda(B \mid C_m) - \lambda(B \mid C_n)$ have the same sign and are non-zero, so

$$\big(\lambda(A \mid C_m) - \lambda(A \mid C_n)\big)\big(\lambda(B \mid C_m) - \lambda(B \mid C_n)\big) > 0 \ (m \neq n)$$

which means that $C$ satisfies the statistical relevance condition.

We have shown that in the space $\langle [0, 1), W, \lambda \rangle$ the countably infinite set $C$ is an SCCS for $\langle A, B \rangle$. □

## 2.1 The "Deductive" Explanatory Feature

We have already said that one of the two explanatory features of statistical common causes considered by Reichenbach was the "deductive" one; namely, from the fact that the conjunctive fork conditions are satisfied for $A$, $C$ and $B$ one can deduce the correlation between $A$ and $B$. We have to justify not reproducing here the original proof by Reichenbach, which was simple and to the point, and using a bit more complicated argument instead. The reason is that we want to use the following fact, which is a characterization of correlation between two events in terms of their relations to a third event; the relations being "how close the event is to being a screener-off for the pair" and "how statistically relevant it is for both events". The fact makes the deductive feature of statistical common causes (SCCs) evident and will also be useful in various endeavours below.

**Fact 1.** *If events A and B are correlated, that is,*

$$P(AB) > P(A)P(B),$$

*then for all events C such that* $0 < P(C) < 1$

$$\frac{P(AB|C) - P(A|C)P(B|C)}{P(C^{\perp})} + \frac{P(AB|C^{\perp}) - P(A|C^{\perp})P(B|C^{\perp})}{P(C)} > \\ - [P(A|C) - P(A|C^{\perp})][P(B|C) - P(B|C^{\perp})]. \quad (2.1)$$

*Conversely, if there exists an event C such that (2.1) is satisfied, then events A and B are correlated.*

*Proof.* Assume that $C$ is such that $0 < P(C) < 1$. Write $P(AB)$ as $P(AB|C)P(C) + P(AB|C^{\perp})P(C^{\perp})$, $P(A)$ as $P(A|C)P(C) + P(A|C^{\perp})P(C^{\perp})$ and $P(B)$ as $P(B|C)P(C) + P(B|C^{\perp})P(C^{\perp})$. After straightforward calculations we see that

$$\begin{aligned} P(AB) - P(A)P(B) = & P(C)P(C^{\perp})\big[ - P(A|C)P(B|C^{\perp}) - P(A|C^{\perp})P(B|C)\big] + \\ & + P(AB|C)P(C) - P(A|C)P(B|C)\big[P(C)\big]^2 + \\ & + P(AB|C^{\perp})P(C^{\perp}) - P(A|C^{\perp})P(B|C^{\perp})\big[P(C^{\perp})\big]^2. \end{aligned} \quad (2.2)$$

Considering the last two lines of the above equation, observe that

$$P(AB|C)P(C) - P(A|C)P(B|C)\big[P(C)\big]^2 = \\ P(C)P(C^{\perp})\Big[\frac{P(AB|C) - P(A|C)P(B|C)}{P(C^{\perp})} + P(A|C)P(B|C)\Big]$$

and

$$P(AB|C^{\perp})P(C^{\perp}) - P(A|C^{\perp})P(B|C^{\perp})[P(C^{\perp})]^2 = P(C)P(C^{\perp})\Big[\frac{P(AB|C^{\perp}) - P(A|C^{\perp})P(B|C^{\perp})}{P(C)} + P(A|C^{\perp})P(B|C^{\perp})\Big].$$

After substituting the two above expressions in equation (2.2) we can infer that

$$P(AB) > P(A)P(B) \equiv \Big(\frac{P(AB|C) - P(A|C)P(B|C)}{P(C^{\perp})} + \frac{P(AB|C^{\perp}) - P(A|C^{\perp})P(B|C^{\perp})}{P(C)} > - [P(A|C) - P(A|C^{\perp})][P(B|C) - P(B|C^{\perp})]\Big).$$

Therefore, if $A$ and $B$ are correlated, (2.1) is valid for any $C$ of non-zero and non-one probability. In the other direction, if we find *some* event $C$ for which (2.1) is valid, due to the above equivalence we can deduce the correlation of $A$ and $B$. □

It is immediate from inspecting the form of (2.1) that if $C$ is a statistical common cause of $A$ and $B$, then (2.1) holds. We can thus claim the following (which was originally proven in Reichenbach (1971)):

**Corollary 2.** *Let $\langle \Omega, \mathfrak{F}, P \rangle$ be a probability space. Let $A, B \in \mathfrak{F}$. If $C \in \mathfrak{F}$ is a statistical common cause of $A$ and $B$ then $A$ and $B$ are correlated.*

A similar result for statistical common cause systems is due to Hofer-Szabó and Rédei (2004):

**Fact 3** (Hofer-Szabó and Rédei (2004))**.** *Let $\langle \Omega, \mathfrak{F}, P \rangle$ be a probability space. Let $A, B \in \mathfrak{F}$. If $\mathbf{C} \subseteq \mathfrak{F}$ is a statistical common cause system of $A$ and $B$, then $A$ and $B$ are correlated.*

## 2.2 In Search for Common Causes, Screening Off Is Enough

Let us now change the perspective. Suppose we know that some events $A$ and $B$ are correlated and we are looking for a statistical common cause. It turns out that it is enough to find an event $C$ such that both it and its complement screen off $A$ and $B$; it is then guaranteed that either $C$ or $C^{\perp}$ will be an SCC for the correlated events. This is because in such a situation the left-hand side of inequality (2.1) will be 0, therefore $[P(A|C) - P(A|C^{\perp})][P(B|C) - P(B|C^{\perp})]$ will be positive, which means that both differences have the same sign—so either $C$ or $C^{\perp}$ will meet the conditions for being a statistical common cause for $A$ and $B$. Let us summarize this in a corollary, for future reference in chapter 5:

**Corollary 4.** *Let $\langle \Omega, \mathcal{F}, P \rangle$ be a probability space. Let $A, B, C \in \mathcal{F}$. Suppose A and B are correlated. If both C and $C^{\perp}$ screen off A from B, then either C or $C^{\perp}$ is a statistical common cause of A and B.*

In other words, if—given two correlated events—we find a two-element partition of unity of the space such that both its elements screen off the correlation, we are guaranteed that (1) one of the elements is a statistical common cause and (2) one of the elements is positively statistically relevant for both correlated events, which is a somewhat intuitive feature of probabilistic causes (this being one of the biggest foundational issues of probabilistic causality). We separate (1) from (2), because they can come apart when we switch our attention to more than two-element partitions of unity which consist of screener-offs only. The following examples will show that, in general, if such a partition of unity (which we will call a *screener system*, see the definition below) has more than 2 elements, we can neither infer that we have found an SCCS, nor that at least one of its elements is positively statistically relevent for both $A$ and $B$.

**Definition 11 [Screener System]** Let $\langle \Omega, \mathcal{F}, P \rangle$ be a probability space. Let $A, B \in \mathcal{F}$. If $\mathcal{C}$ is a partition of unity of $\mathcal{F}$ then $\mathcal{C}$ is called a *screener system* for $A$ and $B$ if all elements of $\mathcal{C}$ screen off $A$ from $B$.

We already know that a two-element screener system for correlated $A$ and $B$ has to contain a statistical common cause for the two events. Both examples to follow will present, for given correlated events $A$ and $B$, three-element screener systems which will *not* be statistical common cause systems.

**Example 1** Consider a probability space $\langle \Omega, \mathcal{F}, P \rangle$, where $\Omega = \{1, \ldots, 100\} \subset \mathbb{N}$, $\mathcal{F} = \Omega^2$, and $P$ is the uniform measure on $\mathcal{F}$ (for any $x \in \Omega$, $P(\{x\}) = \frac{1}{100}$). Consider events $A = \{1, \ldots, 30\}$ and $B = \{11, \ldots, 40\} \cup \{51, \ldots, 85\}$. $P(AB) = \frac{2}{10} > \frac{195}{1000} = P(A)P(B)$, so $A$ and $B$ are correlated. Define $\mathcal{C} := \{C_i\}_{i \in \{0,1,2\}}$, where $C_0 := \{11, \ldots, 20\}$, $C_1 := \{1, \ldots 10\} \cup \{21, \ldots, 50\}$, and $C_2 := \{51, \ldots, 100\}$. Check that

$$
\begin{aligned}
P(AB|C_0) &= \; 1 &&= P(A|C_0)P(B|C_0);\\
P(AB|C_1) &= \; \tfrac{1}{4} &&= P(A|C_1)P(B|C_1);\\
P(AB|C_2) &= \; 0 &&= P(A|C_2)P(B|C_2),
\end{aligned}
$$

therefore $\mathcal{C}$ is a screener system for $A$ and $B$. However, $P(A|C_2) = 0 < P(A|C_1)$, but $P(B|C_2) = \frac{7}{10} > \frac{1}{2} = P(B|C_1)$, which means that $\mathcal{C}$ does not satisfy the statistical relevance conditions for $\{A, B\}$ (see definition 7, p. 5), and so is not a statistical common cause system for $A$ and $B$.

In the above example both $C_0$ and $C_1$ were positively statistically relevant for both $A$ and $B$. We will now see that it may happen that *no* element of a screener system is positively statistically relevant for any of the two correlated events.

**Example 2** Consider a probability space $\langle \Omega, \mathcal{F}, P \rangle$, where $\Omega = \{1, \ldots, 6\} \subseteq \mathbb{N}, \mathcal{F} = \Omega^2$, and $P$ is the uniform measure on $\mathcal{F}$ (for any $x \in \Omega$, $P(\{x\}) = \frac{1}{6}$). Consider events $A = \{1, 2, 3\}$ and $B = \{2, 3, 4\}$. $P(AB) = \frac{1}{3} > \frac{1}{4} = P(A)P(B)$, so $A$ and $B$ are correlated. Define $\mathcal{C} := \{C_i\}_{i \in \{0,1,2\}}$, where $C_0 := \{1, 2\}$, $C_1 := \{3, 4\}$, and $C_2 := \{5, 6\}$. Check that

$$
\begin{aligned}
P(AB|C_0) &= \tfrac{1}{2} &&= P(A|C_0)P(B|C_0);\\
P(AB|C_1) &= \tfrac{1}{2} &&= P(A|C_1)P(B|C_1);\\
P(AB|C_2) &= 0 &&= P(A|C_2)P(B|C_2),
\end{aligned}
$$

therefore $\mathcal{C}$ is a screener system for $A$ and $B$. Notice, however, that $P(B|C_0) = \frac{1}{2} = P(B|C_0^{\perp})$, $P(A|C_1) = \frac{1}{2} = P(A|C_1^{\perp})$, $P(A|C_2) < P(A|C_2^{\perp})$ and $P(B|C_2) < P(B|C_2^{\perp})$. Therefore none of the elements of $\mathcal{C}$ is positively statistically relevant for both $A$ and $B$—even though $C_0$ raises the probability of $A$ and $C_1$ raises the probability of $B$.

A part of the motivation for calling a statistical common cause a "cause" in the first place is that it raises the probability of both its "effects". We now see that, in general, in case of more than two-element screener systems we do not have the guarantee that one of its elements will do this job. On the other hand, not all elements of a finite screener system for two correlated events $A$ and $B$ may *lower* the probability of both $A$ and $B$ (we conjecture that the fact holds also in the case of infinite systems):

**Fact 5.** *Let $\langle \Omega, \mathcal{F}, P \rangle$ be a probability space. Let $A, B$ be correlated events in $\mathcal{F}$. Suppose a finite $\mathcal{C} = \{C_i\}_{i \in I}$ is a screener system for $A$ and $B$. Then, for some $i \in I$, $P(A|C_i) > P(A)$ or $P(B|C_i) > P(B)$.*

*Proof.* Suppose to the contrary, that for all $i \in I$, $P(A|C_i) \leqslant P(A)$ and $P(B|C_i) \leqslant P(B)$. We know from our assumption that $P(AB) > P(A)P(B)$. But, since $\mathcal{C}$ is a partition of unity of $\mathcal{F}$,

$$
\begin{aligned}
P(AB) = \sum_{i \in I} P(AB|C_i)P(C_i) &= \sum_{i \in I} P(A|C_i)P(B|C_i)P(C_i)\\
&\leqslant P(A)P(B)\underbrace{\sum_{i \in I} P(C_i)}_{1} = P(A)P(B),
\end{aligned}
$$

therefore we arrive at a contradiction. □

## 2.3 "Common Common" Constructs

It is one thing to look for a common cause of a single correlation; it is another to ask whether two (or more) correlations can be explained by means of the same common cause. These issues are important for the discussion of the relationship between the PCC and Bell's inequalities, the topic of chapter 3. In this section we will give formal definitions of common statistical common causes and common screener systems, as well as notice a simple fact regarding the existence of common screener systems for finite families of correlations. For completeness we will also include the concept of common statistical common cause systems.

**Definition 12 [Common SCC, Common Screener System, and Common SCCS]** Let $\langle \Omega, \mathcal{F}, P \rangle$ be a probability space. Let $\mathcal{G} \subseteq \mathcal{F}^2$ be a family of pairs of correlated events in $\mathcal{F}$.

- An event $C$ is called a *common statistical common cause* ("CSCC") for $\mathcal{G}$ (or for all pairs in $\mathcal{G}$) if for every $\langle A, B \rangle \in \mathcal{G}$, $C$ is a statistical common cause for $A$ and $B$.
- If $\mathcal{C}$ is a partition of unity of $\mathcal{F}$, then $\mathcal{C}$ is called a *common screener system* ("CSS") for $\mathcal{G}$ if for every $\langle A, B \rangle \in \mathcal{G}$, $C$ is a screener system for $A$ and $B$.
- If $\mathcal{C}$ is a partition of unity of $\mathcal{F}$, then $\mathcal{C}$ is called a *common statistical common cause system* ("CSCCS") for $\mathcal{G}$ if for every $\langle A, B \rangle \in \mathcal{G}$, $C$ is a statistical common cause system for $A$ and $B$.

Since it is nothing unusual for a correlated pair of events not to have a statistical common cause in the given probability space (see chapter 5), it is not surprising that not all pairs of correlated events have common SCCs. Hofer-Szabó et al. (2002) have established necessary conditions for two correlated pairs of events having a common SCC. We will now show that, in any space, a finite family of correlated pairs always has a common screener system. Let us just note that in the case of a finite probability space the problem is trivial—it suffices to construct a partition of unity of the event algebra from all singletons of the sample space elements[2]. However, the screener system constructed this way—while "doing the job"—can be huge. Within the proof of fact 6 we will construct a more efficient screener system.

**Fact 6.** *Let $\langle \Omega, \mathcal{F}, P \rangle$ be a probability space. Let $\mathcal{G} = \{\{A_i, B_i\}\}_{i \in I} \subseteq \mathcal{F}^2$ be a finite family of pairs of correlated events in $\mathcal{F}$. There exists a partition $\mathcal{C}$ of the unity of $\mathcal{F}$ which is a common screener system for $\mathcal{G}$.*

2 If any of these singletons have zero probability, simply append them to arbitrary singletons with non-zero probability.

*Proof.* Consider the set of atoms of the Boolean subalgebra of $\mathcal{F}$ generated by $\cup\mathcal{G}$. Remove any empty atoms. Add any zero-measure atoms to an arbitrary atom with non-zero measure. Notice that the resulting set $\mathcal{C}$ is a partition of unity of $\mathcal{F}$. It is immediate that $\forall_{i\in I}\forall C \in \mathcal{C}$ either

$$P(A_iB_i|C) = 1 = P(A_i|C)P(B_i|C)$$

or

$$P(A_iB_i|C) = 0 = P(A_i|C)P(B_i|C)$$

(all probabilities are defined, because due to our precautions all elements of $\mathcal{C}$ have positive measure). Therefore, every member of $\mathcal{C}$ screens off all the correlations belonging to $\mathcal{G}$. We conclude that $\mathcal{C}$ is a common screener system for $\mathcal{G}$. □

**Problem 1.** *Do infinite families of correlations also always have common screener systems?*

## 2.4 Explanation *via* Screeners—the General Picture

In this section we will discuss constructs different from SCC(S)s, but which nonetheless share the previously mentioned deductive explanatory feature. Let us first put some labels on the already introduced conditions, for future reference:

$$P(AB|C) = P(A|C)P(B|C) \qquad (\mathrm{SCR}(A,B,C))$$

$$P(A|C) > P(A|C^{\perp}) \qquad (\mathrm{STAT}(A,C))$$

$$P(AB) > P(A)P(B) \qquad (\mathrm{CORR}(A,B))$$

The deductive explanatory feature of common causes postulated by Reichenbach can be expressed as the fact that CORR($A$,$B$) follows from SCR($A$,$B$,$C$), SCR($A$,$B$,$C^{\perp}$), STAT($A$,$C$) and STAT($B$,$C$). In general, let us call a set of conditions *deductively explanatory for* CORR($A$,$B$) if the fact that all conditions from the set are satisfied entails positive correlation between $A$ and $B$. Of course, {CORR($A$,$B$)} is trivially deductively explanatory for CORR(A,B). Reichenbach's conjunctive fork criteria comprise an example of a non-trivial deductively explanatory set of conditions. We already said that screening off alone has some explanatory value tied with the vanishing of the correlation when conditional probability is considered. This brings us to the following problem:

**The screener-off classification problem.** Apart from Reichenbach's criteria for a conjunctive fork, are there any other interesting sets of conditions which would contain SCR($A$,$B$,$C$) and would be deductively explanatory for CORR($A$,$B$)?

The word "interesting" is to mean "non-trivial", to exclude explanations of CORR($A$,$B$) by CORR($A$,$B$) or its equivalents which refer to $A$ and $B$ only. Also, since the overarching goal is to find an explanation for a correlation between two events by means of a third event $C$, the explanatory condition should refer to the third event in a non-trivial way.

From a bit different angle: an event $C$, which is a statistical common cause for $A$ and $B$, can be viewed as an explanation for the correlation between $A$ and $B$ because its existence makes true a certain set of conditions which are deductively explanatory for the correlation. We can ask a question: apart from SCC (and the general concept of SCCS), are there different types of screeners which could act as explanations for the same "deductive" reason?

### 2.4.1 Weakening the Screening Off Condition

Before we provide a general classification of explanatory screeners, let us focus on the "perfectness" of screening off as required by SCR($A$,$B$,$C$). It would be unreasonable in any experimental situation ever to expect the observation of frequencies of any events $A$,$B$,$C$ ("$fr(A)$" and so on) such that $\frac{fr(ABC)}{fr(C)} = \frac{fr(AC)}{fr(C)} \cdot \frac{fr(BC)}{fr(C)}$. Moving now from finite frequencies to probabilities, one could have the idea that a weaker condition than SCR($A$,$B$,$C$), for example the following:

$$\left|P(AB \mid C) - P(A \mid C)P(B \mid C)\right| \leqslant \epsilon, \qquad (\mathrm{SCR}(A,B,C,\epsilon))$$

should also—for some small $\epsilon$—together with STAT($A$,$C$), STAT($B$,$C$) and SCR($A$,$B$,$C^{\perp}$) (or SCR($A$,$B$,$C^{\perp}$, $\epsilon$)) form a set of conditions deductively explanatory for the correlation between $A$ and $B$. In other words, one could consider weakening the notion of a statistical common cause by relaxing the requirement of perfect screening off. We will now show that the existence of such a "weakened" statistical common cause no longer permits us in general to deduce the correlation. That is: for arbitrarily small $\epsilon \in (0,1)$, the set $\{$SCR($A$,$B$,$C$,$\epsilon$), SCR($A$,$B$,$C^{\perp}$), STAT($A$,$C$), STAT($B$,$C$)$\}$ is not deductively explanatory for the correlation between $A$ and $B$.[3]

Choose an $\epsilon \in (0,1)$. Let $i$ be the smallest natural number (bigger than 1) such that $\epsilon \geqslant 10^{-i}$. We will construct a probability space and events $A$,$B$,$C$ satisfying SCR($A$,$B$,$C$,$\epsilon$), SCR($A$,$B$,$C^{\perp}$), STAT($A$,$C$), and STAT($B$,$C$), but not CORR($A$,$B$).

Let the sample space $Q$ consist of natural numbers between 1 and $2 \cdot 10^i$ (inclusive). Let the event algebra $S$ be the set of subsets of $Q$. Let the probability measure $P$ assign to each subset $U$ of $S$ the number $\frac{card(U)}{2 \cdot 10^i}$. Events $A$, $B$ and $C$ are defined as follows (square brackets indicate an interval in the set of natural numbers):

- $A := [1, \frac{10^i}{2}] \cup [10^i + 1, (\frac{3}{2}10^i) - 1]$

3 From this it of course follows that the set $\{$SCR($A$,$B$,$C$,$\epsilon$), SCR($A$,$B$,$C^{\perp}$, $\epsilon$), STAT($A$,$C$), STAT($B$,$C$)$\}$ likewise lacks the deductive explanatory feature for CORR($A$,$B$).

- $B := [\frac{10^i}{4} + 1, \frac{3}{4}10^i + 1]$
- $C := [1, 10^i]$.

It is immediate that STAT($A$,$C$), STAT($B$,$C$) and (since $B \cap C^{\perp} = \emptyset$) SCR($A$,$B$,$C^{\perp}$) are satisfied. Notice that $P(AB \mid C) = \frac{1}{4}$ and $P(A \mid C) = \frac{1}{2}$. $P(B \mid C)$ is equal to $\frac{1}{2} \cdot \frac{10^i+2}{10^i}$. Therefore

$$P(A \mid C)P(B \mid C) - P(AB \mid C) = \frac{1}{4}\left(1 + \frac{2}{10^i} - 1\right) = \frac{1}{2 \cdot 10^i} < 10^{-i} \leqslant \epsilon,$$

so SCR($A$,$B$,$C$,$\epsilon$) is satisfied, too.

But the events $A$ and $B$ are not positively correlated. This can be seen from the fact that while $P(AB) = \frac{1}{8}$,

$$P(A)P(B) = \frac{10^i - 1}{2 \cdot 10^i} \cdot \frac{10^i + 2}{4 \cdot 10^i} = \frac{1}{8} \cdot \frac{(10^i - 1)(10^i + 2)}{(10^i)^2} > \frac{1}{8},$$

which concludes the argument that for arbitrarily small $\epsilon \in (0, 1)$, the set comprised of SCR($A$,$B$,$C$,$\epsilon$), SCR($A$,$B$,$C^{\perp}$), STAT($A$,$C$), and STAT($B$,$C$) is not deductively explanatory for the correlation between $A$ and $B$. To deduce a correlation from a statistical common cause one needs an exact by-the-book SCC—one cannot accept any substitutes which "almost" screen off the effects, even if the difference in probabilities between $P(AB \mid C)$ and $P(A \mid C)P(B \mid C)$ is less then .0000001.

Another example will demonstrate that, supposing a pair of correlated events $A$ and $B$ is considered, if the requirement of screening off is weakened by an arbitrarily small $\epsilon$, we can no longer benefit from the feature discussed in section 2.2, namely: the "almost-perfect" screener $C$ may increase the probability of $A$ and decrease the probability of $B$.

Again, choose an $\epsilon \in (0, 1)$. Let $i$ be the smallest natural number such that $\epsilon \geqslant 10^{-i}$. We will construct a probability space and events $A$,$B$,$C$ satisfying SCR($A$,$B$,$C$,$\epsilon$), SCR($A$,$B$,$C^{\perp}$), STAT($B$,$C$), CORR($A$,$B$), but not STAT($A$,$C$).

Let the sample space $Q$ and measure $P$ be defined exactly as in the last example. Events $A$, $B$ and $C$ are defined as follows (square brackets indicate an interval in the set of natural numbers):

- $A := [1, \frac{10^i}{2}] \cup [10^i + 1, 2 \cdot 10^i]$
- $B := [\frac{10^i}{4}, \frac{3}{4}10^i] \cup [10^i + 1, \frac{3}{2}10^i]$
- $C := [1, 10^i]$.

First, notice that $P(A|C^{\perp}) = 1 > \frac{1}{2} = P(A|C)$, so $C$ decreases the probability of $A$ and STAT($A$,$C$) is violated. On the other hand, SCR($A$,$B$,$C^{\perp}$) is satisfied since $P(A|C^{\perp}) = 1$. Notice that

$$P(B|C^{\perp}) = \frac{1}{2} < \frac{1}{2} \cdot \frac{10^i + \frac{1}{2}}{10^i} = P(B|C),$$

so STAT($B$,$C$) holds. In a similar fashion,

$$P(AB) = \frac{(\frac{3}{4}\cdot 10^i)+1}{2\cdot 10^i} > \frac{3}{4}\cdot\frac{10^i+1}{2\cdot 10^i} = P(A)P(B),$$

so the events $A$ and $B$ are correlated. Lastly, regarding the weakening of screening off:

$$\left|P(AB|C) - P(A|C)P(B|C)\right| = \frac{(\frac{1}{4}\cdot 10^i)+1}{10^i} - \frac{1}{2}\cdot\frac{(\frac{1}{2}\cdot 10^i)+1}{10^i} = \frac{1}{2\cdot 10^i} < 10^{-i} \leqslant \epsilon.$$

This concludes the argument that when looking for a "weakened" statistical common cause (i.e. an event $C$ which would be an SCC for $A$ and $B$ had it not violated SCR($A$,$B$,$C$) by a small margin) for two correlated events $A$ and $B$, weakened screening off is not enough. The statistical relevance conditions have to be checked independently.

### 2.4.2 Introducing *Deductive Explanantes*

Let us now move to the task of classifying explanatory screeners. The tool we will use is inequality (2.1) (p. 33).

Suppose, first, that an event $C$ is found such that both SCR($A$,$B$,$C$) and SCR($A$,$B$,$C^\perp$). When is CORR($A$,$B$) deducible? One look at inequality 2.1 is enough to convince us this is the case if and only $C$ is a statistical common cause or a complement of a statistical common cause.

The case of screeners whose negation is not a screener for the given pair of events is (marginally) more complex. Suppose, then, that an event $C$ is found which screens off $A$ from $B$: SCR($A$,$B$,$C$) holds. What else would have to be true in order for us to be able to deduce CORR($A$,$B$)? After again consulting (2.1), we are left with two main possibilities, depending on whether $C$ behaves "symmetrically" towards the probabilities of $A$ and $B$: whether it increases (or decreases) them both, or increases one but decreases the other. For clarity, we isolate the case in which $C$ does not influence the probability of $A$ or $B$ as a separate one. To sum up:

1. If $C$ (which screens off $A$ from $B$) behaves symmetrically towards the probabilities of $A$ and $B$, then from its existence we can infer CORR($A$,$B$) only when

$$[P(A|C) - P(A|C^\perp)][P(B|C) - P(B|C^\perp)] > \frac{P(A|C^\perp)P(B|C^\perp) - P(AB|C^\perp)}{P(C)} \quad (2.3)$$

2. If $C$ (which screens off $A$ from $B$) behaves asymmetrically towards the probabilities of $A$ and $B$, then from its existence we can infer CORR($A$,$B$) only when

$$\frac{P(AB|C^\perp) - P(A|C^\perp)P(B|C^\perp)}{P(C)} > [P(A|C) - P(A|C^\perp)][P(B|C) - P(B|C^\perp)] \quad (2.4)$$

3. If $C$ (which screens off $A$ from $B$) does not change the probability of $A$ or $B$ ($P(A|C) = P(A|C^\perp)$ or $P(B|C) = P(B|C^\perp)$), then from its existence we can infer CORR($A$,$B$) only if

$$P(AB|C^\perp) - P(A|C^\perp)P(B|C^\perp) > 0. \quad (2.5)$$

We propose to call an event $C$ which screens off event $A$ from $B$ a *deductive explanans* of the correlation between $A$ and $B$ if it is either a statistical common cause of $A$ and $B$, or the complement of one, or meets any of the three conditions in the above list. This definition captures the notion of an event whose discovery contributes to the explanation of a given correlation not only by means of screening off but also by means of the previously discussed deductive feature. From the above considerations it should also be evident that no other type of screeners possessing the deductive feature exists; this boils down again to inequality (2.1).

*Deductive explanantes* may be useful in describing situations in which many partial common causes of a correlation are involved. Reichenbach advised giving attention to the disjunction of all such causes[4]; on the other hand, it may well be that even one of them, considered in isolation, has explanatory value for the given correlation.

Consider an example in which little Jimmy is dragged by his mother to a philharmonic hall for his first experience with classical music. His aunts Ann ($A$) and Betty ($B$) play the first violin. To his amazement, Jimmy notices a perfect correlation between the movements of his aunts: for example, whenever Annie enters, Betty likewise begins playing. This amazing coincidence is explained when Jimmy widens his attention to include more of the stage and notices a person furiously waving his hands at the orchestra (the conductor ($C$)). Since the players in the first violin section are well-trained, the appropriate gesture of the conductor is always a signal clear enough to make Ann start playing; the fact that Betty begins to play, too, is irrelevant.[5] This expresses the idea of SCR($A$,$B$,$C$). It is evident that STAT($A$,$C$) and STAT($B$,$C$) hold, too—a conductor's gesture at a given moment increases the probability that the players will enter shortly after. However, suppose the conductor has a bad day and forgets to clearly point out all the entrances; perhaps the music is so complicated he had earlier made a deliberate selection of important entrances he would like to stress. But even if he does not point out a 1st violin entrance, Ann and Betty (typically) also begin to play at the same time, therefore SCR($A$,$B$,$C^{\perp}$) does not hold! Why is that? Because they have notes lying on their desk ($D$). $C$ and $D$ are both partial common causes of the correlation between $A$ and $B$.

$C$ fails to be a statistical common cause for $A$ and $B$, but this does not mean it should be dismissed as an explanation. To the contrary, we would intuitively say that the correlation between entrances of the 1st violin players is explained by the movements of the conductor. And thus it is fortuitous that $C$ is a *deductive explanans* for the correlation (case 1. in the list above; the left-hand side of the inequality is positive while the right-hand side is negative).

4 See p. 159 of Reichenbach (1971).

5 This example could easily be made formal by quantization of time; e.g. choosing the duration of a sixty-fourth note as the time unit.

In chapter 5 we will prove that in finite probability spaces with the uniform measure every correlation between logically independent events has a *deductive explanans*. However, we have to note that—since the conditions for a *deductive explanans* are weaker than those for an SCC—it is "easier" for an event to be a *deductive explanans* than it is for an event to be a statistical common cause. This means that there will be more "false positives", i.e. events which meet the probabilistic requirements from the definition of *deductive explanans* but in fact fail to be genuinely explanatory for the given correlation despite screening it off and possessing the deductive explanatory feature.

# 3 The Principle of the Common Cause and the Bell Inequalities

In this chapter we will discuss the famous Bell inequalities and whether the fact they are falsified both by the predictions of quantum mechanics and by empirical tests impugns the Principle of the Common Cause, and, if so, which version of it is in danger.

## 3.1 Towards the Derivations of the Inequalities

### 3.1.1 Introduction

Consider a source emitting pairs of spin-$\frac{1}{2}$ particles prepared in the singlet state $\frac{1}{\sqrt{2}}(|\uparrow\downarrow\rangle - |\downarrow\uparrow\rangle)$. Assume that each particle travels towards one of two spatially separated detectors (which we will label "L" and "R"). During the flight of the particles each detector is set to measure the spin of the particle in a certain direction. The detectors are situated so that light emitted on measurement at one detector cannot reach the location of the remaining detector before the other measurement takes place. Assume there is a finite set of possible detector settings. The result of the measurement is always binary—"up" or "down"[1], which we will refer to by "+" and "–". From the formalism of quantum mechanics it follows that the probability of obtaining a "+" result on both particles is equal to $\frac{1}{2}\sin^2(\frac{\phi_{ij}}{2})$, where $\phi_{ij}$ is the angle between the direction $i$, set on the left detector, and direction $j$, set on the right detector. And so, if the directions are identical, a perfect anticorrelation of results is expected—a "+" from the left detector means a "–" from the right one. On the other hand, the probability of obtaining a "+" from a detector is predicted to be $\frac{1}{2}$, regardless of the setting. Since the joint probabilities are not, in general, equal to $\frac{1}{4}$, there will be correlations between the results. The stunning result of Bell (1964) is that if a hidden variable (e.g. the complete state of the source) is posited as screening off the results, it is possible (with some additional intuitive assumptions) to derive inequalities falsified by the above predictions. This has been subsequently corroborated experimentally (see the classical paper Aspect et al. (1982) or, for newer results, Scheidl et al. (2010)).

The exposition in the last paragraph was necessarily informal, since we did not want to settle in advance the formalism in which the Bell inequalities are to be formulated and discussed. This is due to the fact that there are two approaches present in the literature, frequently called "big space" and "many spaces approach". Since the

1 We do not include particles which are emitted by the source but do not hit any of the detectors in the picture. There are models for Bell-type correlations which exploit the inefficiency of detectors—see section 3.3.

PCC in various formulations considers the existence of events in probability spaces, to judge the force with which the violation of Bell inequalities strikes the PCC we have to be clear how the probability spaces involved look like.

### 3.1.2 The Big Space Approach and the Many Spaces Approach

Consider a probability space $\langle \Omega, \mathcal{F}, P \rangle$. Any event $A \in \mathcal{F}$ with non-trivial probability[2] induces a new measure on the same event space: for any $C \in \mathcal{F}$, $P_A(C) := P(C|A)$. In fact, a "smaller" probability space is induced: $\langle A, \mathcal{F}_A, P_A \rangle$, where $\mathcal{F}_A := \{C \cap A | C \in \mathcal{F}\}$. In short, an $n$-element partition of the sample space induces $n$ "smaller" probability spaces, provided that each element of the partition has positive probability. The probability of an event in one of the smaller spaces is interpreted as conditional probability in the original "big" space.

In the other direction, suppose you have two probability spaces $\langle \Omega_1, \mathcal{F}_1, P_1 \rangle$ and $\langle \Omega_2, \mathcal{F}_2, P_2 \rangle$. You can then build a "bigger" probability space. First, take as the new sample space $\Omega$ the Cartesian product of $\Omega_1$ and $\Omega_2$. Then consider the set of "rectangles", that is, sets of the shape $A_1 \times A_2$ for some $A_1 \in \mathcal{F}_1, A_2 \in \mathcal{F}_2$. The first task, before the set of rectangles is expanded to be a proper event space, is to define the measure $P$ on it. And the only requirement is that it should have the so called "marginal property"; that is, for any $A \in \mathcal{F}_1$, $P(A \times \Omega_2)$ should be equal to $P_1(A)$ and similarly $P(\Omega_1 \times B)$ should be equal to $P_2(B)$ for any $B \in \mathcal{F}_2$. It is due to this requirement that each of the smaller spaces is embeddable in the big one. But it is easy to see that there is in general more than one way of defining the measure $P$ on the set of rectangles, and therefore, one cannot speak of "the" big space constructed from the smaller spaces.

A different, more informal, but perhaps more illuminating difficulty concerns assigning weights to alternatives. Suppose you have a fair coin and can either toss it (with probabilities $P_A(H) = P_A(T) = \frac{1}{2}$), or conduct a chemical experiment on it in which the presence of nickel in the coin will be assessed (to the best of your knowledge, the probability is $P_B(N) = \frac{9}{10}$). The choice is yours. The probabilities are given by two measures, $P_A$ and $P_B$. Suppose you would like somehow to combine them into a single measure $P$, saying that $P_A(H)$ is to be understood as "the probability of the coin landing heads given that I decide to toss it", in other words, as $P(H|A)$. The problem is that your new measure has to ascribe probability to the event $A$ itself; in a case such as this one, when occurrence of $A$ depends on your choice, one can consider it unwise to think of $A$ having any probability whatsoever.

It turns out we encounter a similar problem when describing the Bell-type experiments. From now on, let "$L_i^+$" be the event "the measurement of the spin in direction $i$ of the particle hitting the left detector yielded the result 'up"'. The quantum mechan-

2 We use this term as meaning "different from both 0 and 1".

ical probabilistic algorithm yields numbers naturally interpreted as probabilities in small spaces labeled by the directions of spin measurement chosen at both (or just one) detectors. For example, $P_{13}(L_1^+ \wedge R_3^+)$ is the probability of obtaining two "up" results at detectors set to direction 1 (the left one) and 3 (the right one); as said above, this probability is equal to $\frac{1}{2}\sin^2(\frac{\phi_{13}}{2})$. $P_1(L_1^+)$ is the probability of getting the "up" result at the left detector set to direction 1. Describing the experiment in this way is called the "many spaces approach". It employs as many probability spaces as there are possible combinations of the directions to be chosen at both detectors. However, one could prefer to have a single probability space and instead of writing "$P_{ij}(L_i^+ \wedge R_j^+)$", write "$P(L_i^+ \wedge R_j^+ \mid L_i \wedge R_j)$", where $L_i$ is the event that the direction $i$ has been chosen at the left detector, and similarly for $R_j$. This—the "big space approach"—is frequently encountered in the literature regarding the connection between the PCC and the Bell inequalities (see e.g. van Fraassen (1982) or Hofer-Szabó (2008)). It, however, requires ascribing probabilities to choices of detector setting. This is one of the reasons for which we prefer to start with the small space approach (the other being its naturalness given QM's predictions) and will be using it in the next section.

### 3.1.3 Deriving the Bell Inequalities

Suppose from now on that there are four directions of spin measurement available at both detectors; let them belong to $I = \{1, \ldots, 4\}$. Consider the set of the values of the hidden variable to be $\{\lambda_k\}_{k\in K}$ for some $K$. The first assumption is frequently labeled as "No Conspiracy" ("NC"; it may be also referred to e.g. by "Hidden Autonomy"): the value of the hidden variable should not be statistically relevant for our choices of detector settings. In the small space approach, this is represented as

$$\forall_{i,j,l,m\in I;k\in K}\ P_{ij}(\lambda_k) = P_i(\lambda_k) = P_j(\lambda_k) = P_{lm}(\lambda_k) \qquad \text{(NC)}$$

It would also be unreasonable to think that, given the value of the hidden variable, the direction chosen by us at one detector should be statistically relevant for the result of the measurement conducted at the other detector. This condition is called "Parameter Independence" ("PI"; sometimes labeled e.g. "Hidden Locality"):

$$\begin{aligned}\forall_{k\in K,i,j,l\in I,j\neq l}\ \ & P_{ij}(L_i^+|\lambda_k) = P_{il}(L_i^+|\lambda_k) = P_i(L_i^+|\lambda_k)\\ & P_{ij}(R_j^+|\lambda_k) = P_{lj}(R_j^+|\lambda_k) = P_j(R_j^+|\lambda_k) \qquad \text{(PI)}\end{aligned}$$

(similarly for "down" results).

The last assumption at least partly shares the motivation with PI: given the value of the hidden variable, the result of the measurement at one detector should be statistically irrelevant to the result of the measurement conducted at the other detector.

This condition is called "Outcome Independence" ("OI"):

$$\forall_{k\in K,i,j\in I}\ P_{ij}(L_i^+|\lambda_k \wedge R_j^+) = P_{ij}(L_i^+|\lambda_k)$$
$$P_{ij}(L_i^-|\lambda_k \wedge R_j^-) = P_{ij}(L_i^-|\lambda_k)$$

(similarly for both pairs of "mixed" results and with $L$ and $R$ exchanged). Notice that OI states that each value of the hidden variable screens off the results of the experiment:

$$\forall_{k\in K,i,j\in I}\ P_{ij}(L_i^+ \wedge R_j^+|\lambda_k) = P_{ij}(L_i^+|\lambda_k)P_{ij}(R_j^+|\lambda_k) \qquad \text{(OI)}$$

(similarly for all other three pairs of possible results).

PI and OI can be jointly expressed as the following condition[3], known in the literature as "Factorisability":

$$\forall_{k\in K,i,j\in I}\ P_{ij}(L_i^+ \wedge R_j^+|\lambda_k) = P_i(L_i^+|\lambda_k)P_j(R_j^+|\lambda_k) \qquad \text{(Factor.)}$$

It turns out, as we will see, that PI, OI and NC jointly allow the derivation of the inequality

$$-1 \leqslant P_{13}(L_1^+ \wedge R_3^+) + P_{14}(L_1^+ \wedge R_4^+) + P_{24}(L_2^+ \wedge R_4^+) - P_{23}(L_2^+ \wedge R_3^+) - P_1(L_1^+) - P_4(R_4^+) \leqslant 0, \qquad \text{(Bell-CH)}$$

which is falsified when the angles $\phi_{13}$, $\phi_{14}$, $\phi_{24}$ and $\phi_{23}$ are suitably chosen. Consider e.g. $\phi_{13} = \phi_{24} = \frac{3\pi}{4}$, $\phi_{14} = \frac{5\pi}{4}$, $\phi_{23} = \frac{\pi}{4}$ and $\phi_{ij} = 0$ for any $i = j$. In this case we would get $\frac{\sqrt{2}-1}{2} \leqslant 0$, which is clearly false.

In section 3.1.5 we will present a "direct" derivation of the Bell-CH, with an additional parameter referring to a potential weakening of the OI assumption (perhaps some of the values of the hidden variable are not perfect screeners?). In the coming section, though, we will use PI, OI and NC to arrive at the inequality in an indirect way, via a theorem of Fine (1982a).

### 3.1.4 The Bell Inequalities via a Non-empirical Joint Measure

There are several types of Bell inequalities. The Bell-CH inequality above is one of the so called Clauser-Horne inequalities, which refer to two measurement directions at each detector. A. Fine [1982a] proved a theorem to the effect that all inequalities of this type are derivable if and only if there exists a probability distribution over four-tuples of measurement results at all possible detector settings which returns the experimental probabilities as marginals. Such a distribution must of course be non-empirical, since it will ascribe non-zero probabilities to events such as "$L_1^+ \wedge L_2^+ \wedge R_3^+ \wedge R_4^+$", which

3 See Jarrett (1984) for a discussion of this point (which uses different terminology).

are conjunctions of outcomes of measuring incompatible observables. We will show, following Fine's directions (though he used a different formalism in which the role of NC was implicit) how PI, OI and NC permit "gluing" the "small" measures $P_1 \dots P_4$ so that the appropriate "big" measure $P$ is obtained. A similar task was undertaken in Müller and Placek (2001)—however, in the context of branching models. Our considerations will not employ any additional structures.

The following is a corrected and rephrased version of Fine's theorem as presented in Müller and Placek (2001).

**Theorem 2** (Fine (1982a)). *Consider four probability spaces $\mathcal{L}_i$ ($i \in \{1,2\}$; the event algebras $\mathcal{F}_{\mathcal{L}_i}$ have two atoms, $L_i^+$ and $L_i^-$) and $\mathcal{R}_j$ ($j \in \{3,4\}$; the event algebras $\mathcal{F}_{\mathcal{R}_j}$ have two atoms, $R_j^+$ and $R_j^-$). Consider four measures $P_{ij}$ in the joint probability spaces with the sample space consisting of four pairs $\langle L_i^\star, R_j^\star \rangle$ ($L_i^\star \in \{L_i^+, L_i^-\}$, $R_j^\star \in \{R_j^+, R_j^-\}$) and the event algebra being the power set of the sample space. Suppose that for any $i$ and $j$ the measures $P_{ij}$ return $P_i$ and $P_j$ as marginals. Then the following conditions are equivalent:*

- *It is possible to define a joint probability measure $P$ on a sample space consisting of sixteen four-tuples of the shape $\langle L_1^\star, L_2^\star, R_3^\star, R_4^\star \rangle$ ($L_i^\star \in \{L_i^+, L_i^-\}$, $R_j^\star \in \{R_j^+, R_j^-\}$), with the event algebra being the power set of the sample space, in such a way that the measure returns the four joint probabilities $P_{ij}$ as marginals;*
- *The eight given probability measures satisfy the following four* Bell-CH *inequalities*

$$-1 \leqslant P_{ij}(L_i^+ \wedge R_j^+) + P_{ij'}(L_i^+ \wedge R_{j'}^+) + P_{i'j'}(L_{i'}^+ \wedge R_j'^{+}) + \\ -P_{i'j}(L_{i'}^+ \wedge R_j^+) - P_i(L_i^+) - P_{j'}(R_{j'}^+) \leqslant 0, \qquad \text{(Bell-CH)}$$

*for $i, i' \in \{1,2\}$; $j, j' \in \{3,4\}$.*

To improve on clarity, instead of n-tuples and pairs we will write n-element conjunctions.

Consider first the "empirical" measures $P_{ij}$ and $P_i$. Enlarge the corresponding probability spaces so that the atomic events are not measurement results (or pairs of measurement results), but measurement results in conjunction with a value of a hidden variable (e.g., for some $k \in K$, $L_1^+ \wedge \lambda_k$ in $\mathcal{L}_1$ or $L_1^+ \wedge R_3^+ \wedge \lambda_k$ in $\mathcal{L}_{13}$). We can therefore speak e.g. of the probability $P_1(\lambda_k)$ for any $k \in K$. The proposed measure is defined as such,

$$P(L_1^+ \wedge L_2^+ \wedge R_3^+ \wedge R_4^+) = \sum_{k \in K} P_1(L_1^+|\lambda_k) P_2(L_2^+|\lambda_k) P_3(R_3^+|\lambda_k) P_4(R_4^+|\lambda_k) P_1(\lambda_k),$$

and similarly for the remaining four-tuples of possible results; each formula contains $P_1(\lambda_k)$ as its last factor. We will show that if PI, OI and NC are assumed, the measure $P$ returns the experimental probabilities as marginals. It will suffice to consider one case (the reasoning is analogous in other cases); let us show the following:

$$P(L_1^+ \wedge L_2^+ \wedge R_3^+ \wedge R_4^+) + P(L_1^+ \wedge L_2^+ \wedge R_3^+ \wedge R_4^-) + P(L_1^+ \wedge L_2^- \wedge R_3^+ \wedge R_4^+) + \\ + P(L_1^+ \wedge L_2^- \wedge R_3^+ \wedge R_4^-) = P_{13}(L_1^+ \wedge R_3^+). \tag{3.1}$$

The left-hand side of the equality is a sum of non-empirical probabilities, while the right-hand side is the experimental probability of two "up" results given detector settings 1 and 3.

First, notice that due to No Conspiracy we have $P_1(\lambda_k) = P_{13}(\lambda_k)$. Also, by employing Factorisability to each of the four elements of the above sum we can substitute $P_{13}(L_1^+ \wedge R_3^+|\lambda_k)$ for $P_1(L_1^+|\lambda_k)P_3(R_3^+|\lambda_k)$. The left-hand side of (3.1) is then equal to

$$\sum_{k\in K} P_{13}(L_1^+ \wedge R_3^+|\lambda_k)P_{13}(\lambda_k)\Big(P_2(L_2^+|\lambda_k)P_4(R_4^+|\lambda_k) + P_2(L_2^-|\lambda_k)P_4(R_4^+|\lambda_k) + \\ + P_2(L_2^-|\lambda_k)P_4(R_4^+|\lambda_k) + P_2(L_2^-|\lambda_k)P_4(R_4^+|\lambda_k)\Big)$$

which after applying Factorisability to the expression in the big parentheses can be seen to equal

$$\sum_{k\in K} P_{13}(L_1^+ \wedge R_3^+|\lambda_k)P_{13}(\lambda_k) = P_{13}(L_1^+ \wedge R_3^+),$$

as required.

We have shown how adopting PI, OI and NC leads to the Bell inequalities in an indirect way. In the next section we will present a direct derivation.

### 3.1.5 A Bell-CH Inequality from Weakened Screening Off

It should be clear where each of the three assumptions, one of them being the requirement that the values of the hidden variable should screen off the measurement results from each other, was used in the above argument. Is it possible to derive an empirically falsifiable inequality from weaker assumptions—for example, that the values of the hidden variable are imperfect screeners, with a "margin of error" equal to some non-zero $\epsilon$? The answer turns out to be positive, although in this approach the margin is unfortunately close to being negligible.

Notice first that in any probability space, for any events $A$, $B$ and $C$ the maximal possible value of $|P(AB|C) - P(A|C)P(B|C)|$, intuitively understood as the inverse of the degree of "quality" of $C$ as a screener for $A$ and $B$, is $\frac{1}{4}$. The following will be our amended version of OI:

$$\exists\epsilon\ \forall_{k\in K, i,j\in I}\ P_{ij}(L_i^+ \wedge R_j^+|\lambda_k) = P_{ij}(L_i^+|\lambda_k)P_{ij}(R_j^+|\lambda_k) \pm \epsilon \tag{OI'}$$

That is, the $\epsilon$ is the "margin of error" for *all* values of the hidden variable and all correlations. OI' is trivially true and is meant to be understood as a definition of $\epsilon$ in what follows.

The derivation here proceeds using the method from Clauser and Horne (1974). The presentation is similar to the one in Placek (2000), save for introducing the $\epsilon$. The starting point is the following elementary fact:

$$\forall_{u,u',v,v' \in [0,1]} \; -1 \leqslant uv + uv' + u'v' - u'v - u - v' \leqslant 0. \tag{3.2}$$

Now let us make the following substitutions:

$$u := P_{13}(L_1^+|\lambda_k); \qquad u' := P_{23}(L_2^+|\lambda_k);$$
$$v := P_{13}(R_3^+|\lambda_k); \qquad v' := P_{14}(R_4^+|\lambda_k).$$

Observe that, with the above substitutions, due to PI we know that

$$u = P_{14}(L_1^+|\lambda_k); \qquad u' := P_{24}(L_2^+|\lambda_k);$$
$$v = P_{23}(R_3^+|\lambda_k); \qquad v' := P_{24}(R_4^+|\lambda_k).$$

After taking this into account and multiplying all sides of (3.2) by $P_{13}(\lambda_k)$ we get

$$-P_{13}(\lambda_k) \leqslant P_{13}(\lambda_k) \cdot \Big(P_{13}(L_1^+|\lambda_k)P_{13}(R_3^+|\lambda_k) + P_{14}(L_1^+|\lambda_k)P_{14}(R_4^+|\lambda_k) + \\ + P_{24}(L_2^+|\lambda_k)P_{24}(R_4^+|\lambda_k) - P_{23}(L_2^+|\lambda_k)P_{23}(R_3^+|\lambda_k) - P_{13}(L_1^+|\lambda_k) - P_{24}(R_4^+|\lambda_k)\Big) \leqslant 0.$$

It is now time to employ OI', so the term $\epsilon$ is introduced:

$$-P_{13}(\lambda_k) \leqslant P_{13}(\lambda_k) \cdot \Big(P_{13}(L_1^+ \wedge R_3^+|\lambda_k) \pm \epsilon + P_{14}(L_1^+ \wedge R_4^+|\lambda_k) \pm \epsilon + \\ + P_{24}(L_2^+ \wedge R_4^+|\lambda_k) \pm \epsilon - P_{23}(L_2^+ \wedge R_3^+|\lambda_k) \pm \epsilon - P_{13}(L_1^+|\lambda_k) - P_{24}(R_4^+|\lambda_k)\Big) \leqslant 0.$$

We now use NC and multiply all expressions in the big parentheses by $P_{13}(\lambda_k)$:

$$-P_{13}(\lambda_k) \leqslant P_{13}(L_1^+ \wedge R_3^+|\lambda_k)P_{13}(\lambda_k) \pm \epsilon \cdot P_{13}(\lambda_k) + P_{14}(L_1^+ \wedge R_4^+|\lambda_k)P_{14}(\lambda_k) \pm \epsilon \cdot P_{13}(\lambda_k) + \\ + P_{24}(L_2^+ \wedge R_4^+|\lambda_k)P_{24}(\lambda_k) \pm \epsilon \cdot P_{13}(\lambda_k) - P_{23}(L_2^+ \wedge R_3^+|\lambda_k)P_{23}(\lambda_k) \pm \epsilon \cdot P_{13}(\lambda_k) + \\ - P_{13}(L_1^+|\lambda_k)P_{13}(\lambda_k) - P_{24}(R_4^+|\lambda_k)P_{24}(\lambda_k) \leqslant 0$$

which is by the definition of conditional probability equivalent to

$$-P_{13}(\lambda_k) \leqslant P_{13}(L_1^+ \wedge R_3^+ \wedge \lambda_k) \pm \epsilon \cdot P_{13}(\lambda_k) + P_{14}(L_1^+ \wedge R_4^+ \wedge \lambda_k) \pm \epsilon \cdot P_{13}(\lambda_k) + \\ + P_{24}(L_2^+ \wedge R_4^+ \wedge \lambda_k) \pm \epsilon \cdot P_{13}(\lambda_k) - P_{23}(L_2^+ \wedge R_3^+ \wedge \lambda_k) \pm \epsilon \cdot P_{13}(\lambda_k) + \\ - P_{13}(L_1^+ \wedge \lambda_k) - P_{24}(R_4^+ \wedge \lambda_k) \leqslant 0$$

Let us now sum over $k \in K$. We arrive at two inequalities:

$$-1 \leqslant P_{13}(L_1^+ \wedge R_3^+) + P_{14}(L_1^+ \wedge R_4^+) + P_{24}(L_2^+ \wedge R_4^+) - P_{23}(L_2^+ \wedge R_3^+) - P_1(L_1^+) - P_4(R_4^+) - 4\epsilon;$$

$$P_{13}(L_1^+ \wedge R_3^+) + P_{14}(L_1^+ \wedge R_4^+) + P_{24}(L_2^+ \wedge R_4^+) - P_{23}(L_2^+ \wedge R_3^+) - P_1(L_1^+) - P_4(R_4^+) + 4\epsilon \leqslant 0.$$

If $\epsilon = 0$, the outcome is simply the Bell-CH inequality. Notice that if the angles between measurement directions are chosen as noted on p. 47, we get that $\frac{\sqrt{2}-1}{2} + 4\epsilon \leqslant 0$, which is false for $\epsilon < 0.052$. This is the degree to which we can weaken the requirement of screening off present in OI and still derive a falsifiable inequality, keeping the other assumptions intact and not introducing any new assumptions. Admittedly, it is a very modest weakening.

### 3.1.6 Connection with the PCC

Since the Bell inequalities have been shown to be false, it would seem that one (at least) from the three assumptions—NC, PI or OI—must go. Although there are dissenting opinions (e.g. Stapp's arguments against locality), the majority view is that OI is the culprit. Van Fraassen [1982] was apparently the first to claim that the issue was connected with Reichenbach's ideas; the Bell setup is an example of a "conceivable phenomenon in which there is a correlation for which there can exist no common cause". Structurally, the argument proceeds by *reductio*; if a common cause is posited, Bell inequalities follow. And the existence of a common cause is taken to be expressed by OI—but in the big space approach (van Fraassen labels the assumption as "Causality"). Earlier (p. 100), the author generalizes the notion of a common cause to "not just a yes-no event", leaving out the statistical relevance conditions, thus arriving at the notion which we labeled as "screener system". Of course, if in the case of the Bell setup no screener systems exist (given NC and PI, which we do not want to abandon), then *a fortiori* no statistical common cause systems exist, too, which would seem to be against Reichenbach's idea. Since then, however, a number of authors have raised concerns that one cannot disprove Reichenbach's Principle by such arguments, because they show the nonexistence of a single common cause for all correlations (a *common* common cause), which the Principle does not require to exist. The situation in the literature is roughly as follows (since all the authors use their own formalisms, in the following list we will abstain from using any formalism whatsoever):

- Belnap and Szabó (1996) first note the apparent discrepancy in what the violation of Bell inequalities is taken to prove and what the PCC actually claims; an argument regarding the nonexistence of a common common cause of the Einstein-Podolski-Rosen (EPR) correlations (modally interpreted) is given in the Branching Space-Times setting;[4]
- Szabó (2000) presents a model for the EPR correlations in which different correlations are screened off by different common causes; however, as the author himself

4 The Branching Space-Times (BST) approach to Bell-type experiments is still being developed, see e.g. Placek (2010); however, discussing it here would not be worthwhile, since it would require introducing the BST formalism, and the conclusions do not consider the issue at hand.

notes, the model does not satisfy a stronger (but feasible) version of No Conspiracy. Namely, detector settings may be statistically relevant for some Boolean combinations of the values of common causes; Szabó conjectures that this is inevitable and no "corrected" model can be given;
- Graßhoff et al. (2005) prove Szabó's conjecture by providing a derivation of a Bell-type inequality from *separate* common causes (as opposed to a *common* common cause); however, the derivation is made under the assumption that measurement results are perfectly anti-correlated, which is practically unverifiable;
- Hofer-Szabó (2008) observes that under the assumptions of the above mentioned derivation common common causes may be defined, and so the derivation is only reducibly separate-common-causal; the author presents an irreducibly separate-common-causal derivation for "yes"-"no" common causes;
- independently, Portmann and Wüthrich (2007) address the faults of their previous paper and present an irreducible separate-common-causal (with the common causes building up partitions of arbitrary finite size) derivation of a Bell-CH inequality with the requirement that the anticorrelations be *close* to perfect—the upper bound[5] was given as $2.689 \cdot 10^{-5}$;
- Higashi (2008) also presents a separate-common-causal derivation of a Bell-CH inequality, although this time the parameters are probabilities of detector settings (the author works in a kind of a big-space approach);
- Hofer-Szabó (2011) improves on the bound of Portmann and Wüthrich (2007) by providing a derivation of a falsifiable inequality (of the Wigner-type) with the requirement that the anticorrelations may be non-perfect to the margin of $1.73 \cdot 10^{-2}$;
- Hofer-Szabó (2012) produces a separate-common-causal derivation of the Bell-CH inequality with the additional assumption that the screening off factors operate in a deterministic way (i.e., all relevant conditional probabilities are 0 or 1).

And so the move from common common causes to separate common causes has not lead so far to creation of a fully non-conspiratorial model for the Bell-type correlations which would preserve PI. But the task has not also been proven to be impossible, even if a vast majority of people interested in the topic share the conviction that it is. We will explore some more details of this in section 3.2.[6]

5 Meaning: the probability of a "+" result in one wing given a "+" result in the other wing, which in the case of perfect anticorrelations is equal to zero.

6 Suárez (2007) describes a few different kinds of "causal models" which are supposedly able to explain the failure of Factorisability (the author claims that the NC condition is a necessary condition for Factorisability, so he is concerned only with the failure of the latter, not the former). We will not discuss them here. They are to show that the failure of Factorisability does not exclude *any* sort of causal model for the correlations, but the models offered are obviously just "proofs of concept" and exhibit some controversial features, e.g. a past cause influencing both the emission event and the choice of

Let us consider the connection between OI and the various forms of the PCC. The outcome will be a bit different depending on whether we choose the big or small space approach.

**Big space approach.** This is the formulation of OI in this approach:

$$\forall_{i,j\in I,k\in K}P(L_i^+ \wedge R_j^+|L_i \wedge R_j \wedge \lambda_k) = P(L_i^+|L_i \wedge R_j \wedge \lambda_k)P(R_j^+|L_i \wedge R_j \wedge \lambda_k)$$

(and similarly for all other pairs of measurement results).

As said earlier, there is no mention of the statistical relevance conditions. Still, this could at first sight be simply a weakened version of PCC 3, generalized to a more than two-element partition. The starting point of the above mentioned arguments by Belnap, Szabó, Hofer-Szabó and Rédei could then be the observation that OI is presumed to be in the scope of an existential quantifier referring to $\Lambda$, the set of values of the hidden variable, and therefore the order of the quantifiers is in fact $\exists_{\Lambda=\{\lambda_k\}_{k\in K}}\forall_{i,j\in I,k\in K}$, meaning that by assuming OI we in fact assume the existence of a set of screening factors common for all correlations.

The matter looks differently, though. In this formulation screening off is not done by the values of the hidden variables, the various $\lambda_k$. The role of screeners is played by triples consisting of the values of hidden variables together with the choices of measurement settings on both detectors, for example, "$L_1 \wedge R_3 \wedge \lambda_k$" for some $k \in K$. All such triples taken together do, in fact, constitute a partition of the "big" space. But the elements of this partition are screener-offs for *different* pairs of measurement outcomes, depending on which measurement settings go into the particular screener![7] It could be responded that this description is wrong, since what is to be screened in this case are not correlations, but *conditional* correlations. This, however, is simply moving the discussion to the small space approach. Recall that Reichenbach's Principle concerns correlation between events; a statement that for some $A, B, C$, $P(AB|C) > P(A|C)P(B|C)$ (i.e., a statement of correlation of $A$ and $B$ conditional on $C$) does not fall within its scope.

**Small space approach.** Recall the formulation of OI in this approach:

$$\forall_{k\in K,i,j\in I}\ P_{ij}(L_i^+ \wedge R_j^+|\lambda_k) = P_{ij}(L_i^+|\lambda_k)P_{ij}(R_j^+|\lambda_k)$$

(and similarly for all other pairs of measurement results).

In this formulation, the values of the hidden variables *are* screeners for the correlations in question. What is more, the same $\lambda_k$'s screen off various correlations, with different measurement settings. However, they do so in different probability spaces. In each of these, the values of the hidden variable form partitions of the sample space, and all elements of the partition screen off one set of correlations: the one for the

---

detector settings or faster-than-light causation. One of the models is similar to the one from Butterfield (2007) and will be briefly described in section 3.4.

7 For example, $L_1 \wedge R_3 \wedge \lambda_k$ screens off $L_1^+$ from $R_3^-$, while $L_2 \wedge R_4 \wedge \lambda_k$ screens off $L_2^+$ from $R_4^-$.

particular measurement settings. Screening off of correlations under different measurement settings is done in a different probability space, with (at least potentially) a different measure. Therefore, once again, even if we allow for the generalization of Reichenbach's view to more than two-element partitions of sample spaces, the situation here does not fall within the scope of his Principle due to various spaces being required.

The moral of the last two paragraphs is this: it is not the formal notion of the common cause in Reichenbach's sense, even without the statistical relevance condition, but with the requirement of a partition of screeners, which is undermined by the falsifiability of inequalities derived from OI (among other assumptions).

On the other hand, there is an obvious intuitive connection with a less formal (and still Reichenbachian) view, e.g. that of PCC 1. If we simply require that for any correlation there should exist a common cause, and in the case of the Bell-type setups we are looking for common causes among the (properties of) states of the source on emission, then of course OI says more then we require[8]; it posits the same common causes for all correlations. It is worthwhile, then, to distinguish between common common causes and separate common causes in this case. Most of the papers cited on the list on p. 51, while they do refer to Reichenbach's Principle:

- formally do not operate using exactly Reichenbach's notions, since they either work in various probability spaces or consider conditional correlations; but
- informally fully adhere to Reichenbach's view of requiring common causes, *not* common common causes, for correlations.

There is another—general—way of "saving" the PCC, via the notion of "causal completability" (see e.g. Hofer-Szabó et al. (1999)). Informally speaking (for now), if a probability space lacks a common cause for a correlation, it can always be extended to a bigger space, which preserves the measure on all "old" events, but contains a common cause for the previously unexplained correlation. Discussion of this notion and the proof of some results concerning it is one of the main topics of chapter 6.

Lastly, let us note that the perspective outlined at the beginning of this section—namely: that the failure of Bell's inequalities means that at least one of the three assumptions of its derivations must be abandoned—may be misguided. This is because NC, PI and OI are not independent assumptions which stand or fall in isolation. In fact, the whole trio forms a single assumption, since there is an implicit existential quantification over the set of lambdas, the values of the hidden variable. An a bit extreme illustration of the idea would be this: suppose someone says "consider an integer $x$ such that 1) $x \leqslant 0$ and 2) $x > 0$". Of course, there can be no such $x$, because the proposed conditions are mutually exclusive. But we do not conclude that one of them

8 Even if we use the small space approach and adopt PCC 2.

is false and should meet with general abandonment. The situation here is different, because we have intuitive grounds to believe that NC and PI should hold. Still, the failure of the Bell inequalities means that the three conditions cannot jointly hold on the set of values of the posited hidden variable. If it is non-conspiratorial and screens off the measurement results at one wing from detector settings at the other wing, then it cannot be a screener-off for all the correlations between measurement results.

## 3.2 Separate Common Causal Explanations

### 3.2.1 Separate Common Causes—*Contra* and *Pro*

Placek (2009) claimed that in the case of EPR-type correlations the distinction between (separate) common causes and common common causes is a red herring.[9] According to OI, each value $\lambda_k$ of the hidden variable screens off the correlations (the proponents of the big space approach would say "conditional correlations"). This $\lambda_k$ is supposed to be a complete state of the system on emission of the two particles, typically "a different state from the quantum-mechanical pair's state", which "is assumed to be an incomplete state of the pair" (Berkovitz (2008)). Two emitted pairs in the same quantum state may be in different complete states. If this interpretation of various lambdas is adopted, then, were there different screener systems for different correlations, it would follow that a system could simultaneously be in two different complete states—and so the states would not be complete after all. I believe that this strike at the notion of separate common cause explanation hits the target only if we cling to the notion that ultimately, what screens off correlations are complete states, or—in other words—complete descriptions of objects in question. In my opinion we should *not* expect this to happen. I will first give an (abstract) example of how we are inclined to accept screening off by incomplete states and would not expect screening off by (more) complete states.[10]

Look at figure 3.1. Consider a population in which two pairs of symptoms are correlated: symptom $A$ with symptom $B$ and symptom $C$ with symptom $D$. There is no information suggesting direct causation between either $A$ and $B$ or $C$ and $D$; what is more, there is similarly no information regarding causal connections between the two pairs: neither $A$ nor $B$ are thought to be causally relevant for $C$ or $D$, and *vice versa*. Suppose two previously hidden genetic features are discovered (between which there is also no hint of a causal connection), which meet the requirements for a statistical

9 Similar misgivings are cited by Hofer-Szabó (2011) and attributed to an anonymous referee.

10 Of course, formally there is nothing like a "more" complete state, since states are either complete or incomplete, but the gist of the example should be obvious: in general we do not expect that by including more information in the description of the events in question we close in on "real" screeners.

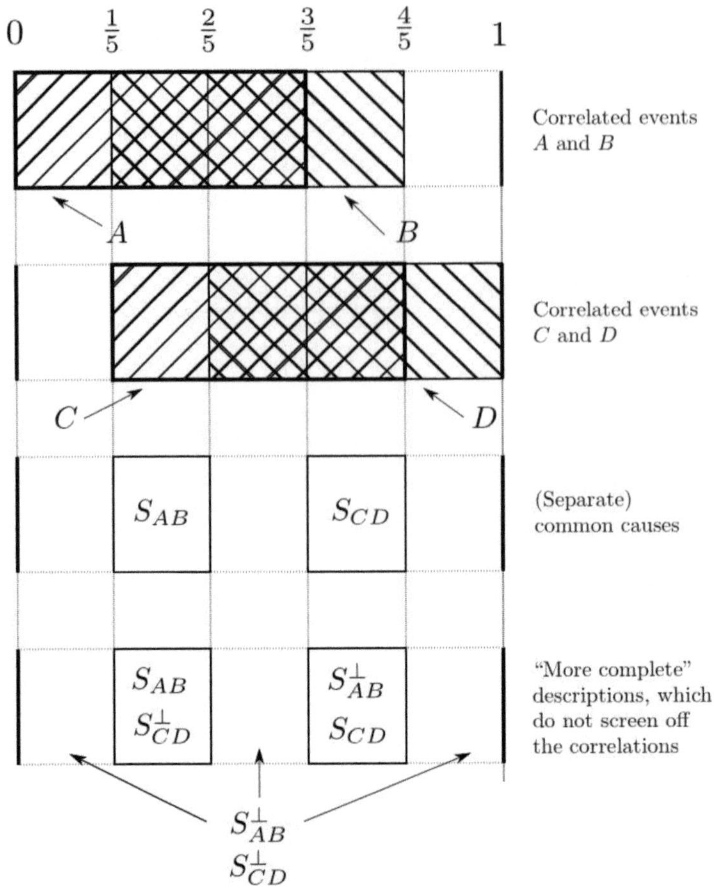


**Figure 3.1:** Two correlations, each screened off by an incomplete state description, but not screened off by some complete descriptions.

common cause from PCC 3: all people with the trait $S_{AB}$ display both symptoms $A$ and $B$, while the absence of the trait $S_{AB}$ makes the display of symptoms $A$ and $B$ statistically independent; similarly, all people with the trait $S_{CD}$ display both symptoms $C$ and $D$, while in the absence of the trait $S_{CD}$ the symptoms $C$ and $D$ are statistically independent. It is natural to conclude that both correlations are explained by their (separate) common causes, $S_{AB}$ in one case, $S_{CD}$ in the other. These are *not* complete descriptions, or complete states; suppose for clarity that other predicates are excluded and the individuals are to be described by means of possession (or lack) of the two genetic traits in question. Then, a complete state would be a Boolean combination of $S_{AB}$ and $S_{CD}$, but we would have no reason at all to expect any such combination to screen off both correlations. In the situation depicted in figure 3.1, the complete state $S^{\perp}_{AB} \cap S^{\perp}_{CD}$ screens off neither $A$ and $B$ nor $C$ and $D$.

We can find a related, less formal example in the philosophy of biology literature, namely that concerning the levels of selection. For example, in Sober (1992) we see that the participants of the discussion consider full phenotypic descriptions to be acting as screeners. These descriptions are not "complete states" in our sense, since there are many genotypical realisations for a given phenotype. The correlations to be screened off are between survivability and possessing a certain genotype. Consider two correlations involving survivability and two genotypes, screened off by two distinct full phenotypic descriptions; in the jargon of this work we would describe the situation as displaying two separate screener systems (see definition 11, p. 35). Now, logical combinations of two distinct full phenotypic descriptions will not in general be full phenotypic descriptions (for the case similar to the one illustrated above, take the conjunction of negations of the two descriptions) and so will not be expected to act as a screening off factor in the correlations under discussion. However, it has to be observed that in this example one of the correlated factors, namely, possession of a certain genotype, is a plausible candidate for being a complete state description and in any case it will trivially screen off survivability from possessing any genotype; therefore the example is not ideal. It is still true, though, that on the level on which the participants of the discussion situate the screening factors, logical combinations of screeners are not in general screeners themselves.

For the remaining part of this section, notwithstanding the lack of a conceptual description of a separate common cause in the EPR context, I take the idea of such to be not *a priori* rejected, and so worth considering. I will finish by stating two open formal problems concerning the issue.

### 3.2.2 Towards a Separate-common-cause Model of EPR Correlations

To recapitulate, the issue is this: is it possible to build a probability space modelling the EPR correlations which would feature four screener systems, one for each pair of detector settings, satisfying Outcome Independence (in the sense that the screener system for a particular pair of detector settings should screen off the correlation arising at these detector settings), Parameter Independence and No Conspiracy (in the sense that no Boolean combination of elements of the screener systems should be correlated with any pair of detector settings)? Such a (hypothetical) probability space will be called a "local non-conspiratorial model with separate screener systems" ("separate-ss model" for short) and will be defined shortly. (Since the literature on this topic uses prevailingly the big space approach, we will adopt said approach in this section.) The question regarding the existence of such models is still open.

Suppose, then, that the role of "separate common causes" is to be played by four partitions $\{\lambda_{ij}^k\}_{k\in K(ij)}$; $K(ij)$ is the index set for the screener system for the pair of detector settings $L_i$ and $R_j$. The conditions of PI, OI, and NC are then expressed as follows[11]:

$$P(L_i^m \mid L_i R_j \lambda_{ij}^k) = P(L_i^m \mid L_i R_{j'} \lambda_{ij}^k) \qquad (\mathrm{PI}^s)$$
$$P(R_j^n \mid L_i R_j \lambda_{ij}^k) = P(R_j^n \mid L_{i'} R_j \lambda_{ij}^k) \qquad (\mathrm{PI}^s)$$
$$P(L_i^m R_j^n \mid L_i R_j \lambda_{ij}^k) = P(L_i^m \mid L_i R_j \lambda_{ij}^k) P(R_j^n \mid L_i R_j \lambda_{ij}^k) \qquad (\mathrm{OI}^s)$$
$$P(L_i R_j \mathfrak{A}) = P(L_i R_j) P(\mathfrak{A}) \qquad (\mathrm{NC}^s)$$

where $\mathfrak{A}$ is any Boolean combination of the elements of four partitions $\{\lambda_{ij}^k\}$ ($i = 1, 2, j = 3, 4$) and all formulas are quantified for all $i \in \{1, 2\}$, $j \in \{3, 4\}$, $m, n \in \{+, -\}$ and $k \in K(ij)$. We will *not* require a screener system for one correlation to satisfy Parameter Independence (and Outcome Independence) with respect to another correlation.

**Definition 13**[**local non-conspiratorial separate-ss (common-ss) models**] Consider a probability space $\langle \Omega, \mathcal{F}, P \rangle$ with events $L_i$ and $R_j$ corresponding to detector settings, and $L_i^m$ and $R_j^n$ corresponding to measurement events ($i \in \{1, 2\}, j \in \{3, 4\}$, $m, n \in \{+, -\}$). Suppose that the model exhibits the EPR correlations.

The probability space is a local non-conspiratorial model with **separate screener systems** (or a **separate-ss model**) for the EPR correlations if there exist four partitions of $\Omega$, $\{\lambda_{ij}^k\}_{k \in K(ij)}$, one for each pair of detector settings, such that for each such a pair $L_i$,$R_j$ the partition $\{\lambda_{ij}^k\}_{k \in K(ij)}$ meets the conditions of Parameter Independence ($\mathrm{PI}^s$), Outcome Independence ($\mathrm{OI}^s$) and No Conspiracy ($\mathrm{NC}^s$) with regard to the correlations arising at the detector settings $L_i$ and $R_j$.

The model is **deterministic** if, in addition, for any $i, j$ and for any $k \in K(ij)$, the probabilities $P(L_i^m | L_i R_j \lambda_{ij}^k)$ and $P(R_j^n | L_i R_j \lambda_{ij}^k)$ are equal to 0 or 1.

The probability space is a local non-conspiratorial model with **a common screener system** (or a **common-ss model**) for the EPR correlations if there exists a single partition of $\Omega$, $\{\lambda^k\}$, which meets the conditions of $\mathrm{OI}^s$, $\mathrm{PI}^s$ and $\mathrm{NC}^s$ with regard to *all* of those correlations.

The model is **deterministic** if, in addition, for any $i, j$ and for any $k$ the probabilities $P(L_i^m | L_i R_j \lambda^k)$ and $P(R_j^n | L_i R_j \lambda^k)$ are equal to 0 or 1.

Notice that a common-ss model is, trivially, a separate-ss model (each correlation has its own screener system and these just happen to coincide).

### 3.2.3 Bell Inequalities as Proof of Nonexistence

The usual strategy of showing that the EPR correlations cannot be modelled in a certain way (e.g. by introducing a single screening off local hidden variable) is to directly

11 Remember that we frequently omit the "$\cap$" sign: e.g. $L_i R_j \lambda_{ij}^k$ stands for $L_i \cap R_j \cap \lambda_{ij}^k$.

derive the Bell inequalities from the probabilistic assumptions true in the purported probability space (in the small space approach we would of course have to be talking about spaces in the plural). I have already mentioned in section 3.1.6 that this feat has not been accomplished so far. But it might be interesting to consider an alternative method, which could be called a "meta-model" strategy.

The strategy goes like this: suppose a set of conditions on a probability space $P$ is presented and we would like to prove that such a space cannot exist; we are, however, unable to derive the Bell inequalities from the assumptions about $P$. We could instead try to show that if $P$ existed, then some other $P'$, also modelling the EPR correlations, would exist, in which the Bell inequalities were derivable. Then, by contraposition, since for correlations violating the Bell inequalities $P'$ does not exist, in such cases $P$ does not exist too.

Now, we already know that no local non-conspiratorial common-ss model for the EPR correlations exists, since the Bell inequalities are derivable from the three assumptions. By the "meta-model" strategy, to disperse any hope of a local non-conspiratorial separate-ss model it would be enough to show that any model of this kind can be transformed into a model of the common-ss kind. Tomasz Placek and me undertook this task in 2011; so far, we have been unable to obtain this result, reaching only two partial results, one of which goes in the opposite direction, namely that of proposing a certain model for the correlations.

The goal was to prove that if there is a local non-conspiratorial separate-ss model for some family of EPR correlations, then there is a local non-conspiratorial common-ss model for the same correlations; in other words, that the switch to separate screener systems does not provide greater generality. We have not achieved this goal; in fact, what we arrived at is a general method for creating non-conspiratorial common-ss models for EPR correlations, which may not satisfy Parameter Independence. One way of looking at this is that, since common-ss models are trivially separate-ss models, it connects nicely with the results from Szabó (2000). With regard to the possibility of models for EPR correlations involving separate screener systems, the situation is as follows:

- Szabó (2000) produces models which satisfy $OI^S$ and $PI^S$, but may fail to satisfy $NC^S$;
- in Placek and Wroński (2011) we produce models which satisfy $OI^S$ and $NC^S$, but may fail to satisfy $PI^S$.[12]

**12** The details of the construction, too big to fit in this chapter, are given in the proof of Lemma 5 in Placek and Wroński (2011). The Lemma itself unfortunately claims that what is shown is the abovementioned transformation of a local non-conspiratorial separate-ss models into local non-conspiratorial common-ss models. This is not true; the proof contains a mistake—the final models may not satisfy $PI^S$. Together with T. Placek I would like to thank Samuel Portmann and Adrian Wüthrich for finding this mistake and discussing it with us.

Perhaps it is possible to somehow combine the two constructions and arrive at a proper local non-conspiratorial separate-ss model; I would like to leave it as problem 2 at the end of this section. Remember that the separate-ss models the construction from Placek and Wroński (2011) produces are in fact common-ss models, so any method aiming at making sure that $PI^s$ is satisfied would also have to change the construction so that "genuinely" separate-ss models would be produced.

Suppose one wanted to go in the opposite direction and disprove the possibility of a local non-conspiratorial separate-ss model for the EPR correlations using the "meta-model" strategy. In light of the already mentioned result from Hofer-Szabó (2012), this could be achieved by showing that any separate-ss model can be transformed into a *deterministic* separate-ss model. Presently, the method of making a given model deterministic is only known for common-ss models:

**Fact 7** (Placek). *Any local non-conspiratorial common-ss model for some family of EPR correlations can be transformed into a local non-conspiratorial deterministic common-ss model.*

(One can prove this fact by modifying the construction from the proof of Lemma 5 from Placek and Wroński (2011); we hope to clarify these matters in a future paper.)

We are, then, left with two open problems, leading to opposite directions:

**Problem 2.** *Provide a method of constructing local non-conspiratorial separate-ss models for the EPR correlations.*

**Problem 3.** *Show that any separate-ss model can be transformed into a deterministic separate-ss model.*

If you want to show that explaining the EPR correlations by means of separate screener systems is possible, tackle problem 2. If you want to disprove this possibility, tackle problem 3. Solving one of those will mean the other is unsolvable.

## 3.3 Exploiting the Detection Loophole

In a given run of any experiment similar to the one described at the beginning of this chapter, it may happen that only one detector "fires", or even none of them does—the detectors are inefficient. This may be attributed simply to random errors of the experimental equipment, but one could also entertain the thought that the inefficiency is due to a hidden property of the emitted particles. This is the central idea behind the so called "Prism models" (dating back to Fine (1982b)).

A model of this kind is given by a meticulous construction in Szabó and Fine (2002). The hidden variable may take one of 48 values. It is deterministic in the sense that each value of the hidden variable predetermines whether the given particle will be detected by the detector and, if this happens, what the measurement result will be. Experimental probabilities are recovered. The main trick is a sort of "unfair sampling", introduced so that the detected particles violate Bell-CH inequalities.

Since the model is deterministic, Factorisability (and *a fortiori* OI) has to hold in it. However, such models are understandably generally considered "ad hoc" (Shimony (2009)) and, with the increasing efficiency of detectors, methods were proposed for closing the detection loophole (see again Shimony (2009)).

## 3.4 Common Causes as Hypersurfaces

Lastly, let us briefly mention another PCC-related option present in the literature. Up to now, both in the big space and small space approach, the supposed common causes for Bell-type correlations were events in probability spaces. Butterfield (2007) describes in detail a view (developed first in Butterfield (1989)) called "Stochastic Einstein Locality" (SEL), in which common causes are factors which determine probabilities of events (while nothing is said about the probabilities of the factors themselves). For an event $E$ (associated with a space-time region), any hypersurface $t$ cutting through $E$'s past light cone $C^-(E)$ determines a probability, $P_t(E)$, of $E$'s occurrence.

The author discusses a few formulations of the SEL idea and provides proofs of their formal relations. In one version, if $E$ and $F$ are space-like related events (none is in either past or future light cone of the other) and $t$ is a hypersurface cutting through $\left(C^-(E) \cup C^-(F)\right) \setminus \left(C^-(E) \cap C^-(F)\right)$, then

$$P_t(E \wedge F) = P_t(E)P_t(F).$$ [13]

The similarity with the screening off condition is obvious; it is just that the screener is in the subscript. SEL is also taken to be violated by violations of Bell inequalities. The conceptual difference between SEL and the PCC is that while the PCC (at least in its more formal shapes, like PCC 3) is plausibly falsified by everyday examples (sea levels vs. bread prices etc.) it takes a Bell-type experimental setting to violate SEL.

## 3.5 Summary

The violation of Bell inequalities has an impact on these formulations of the PCC which require screening off. If the only candidates for common causes in this case are complete states of the source on emission, then—if they are not statistically relevant for the choices of detector settings and they in turn are not statistically relevant for the

13 We omit the subscript referring to a possible world.

measurement results in the other wing—the common causes cannot act as screeners, thus violating the first condition in the definition of a statistical common cause. This is one of the motivations for abandoning the general requirement of screening off from the definition of common causes, even if one would like to preserve Reichenbach's idea that common causes for correlations should exist (i.e. PCC 1)). The motivation is similar to the one given in arguments from conservation principles (see e.g. chapter 6 of Cartwright (1989) and section 1.4.1 of the current essay).

The recent results cited in section 3.1.6 suggest that the additional caution gained by the move from assuming common common causes to assuming separate common causes for the various correlations may not be enough to block derivations of empirically falsifiable Bell-type derivations. But the issue has not been ultimately resolved yet, as evidenced by the two open problems from the end of section 3.2.

# 4 The Principle of the Common Cause and the Causal Markov Condition

A big part of modern probabilistic causality is concerned with the task of causal modeling—that is, deriving causal information from statistical data and depicting causal structures by means of graphs (usually directed acyclic graphs, or "DAGs"; definitions will follow). The graphs are frequently called "Bayesian networks"[1]. We have all heard the slogan that "correlation does not mean causation"; it would be a trivialization, but perhaps an illustrative one, to say that the Bayesian networks project begins with a contraposition of "causation means correlation": "independence means absence of (direct) causation". From statistical data, information about independence (and conditional independence) of variables is gathered, and on that basis a DAG is constructed with the variables as nodes and arrows denoting direct causal relationship (according to one of many algorithms available; see Spirtes et al. (2000) and Williamson (2005) for the "adding arrows" algorithm).

There is a condition, called the "Causal Markov Condition" (henceforth usually "CMC"), which is frequently taken as one which a pair consisting of a DAG and a probability distribution for the values of its nodes should satisfy if the pair is to be admitted as reliably depicting a real causal situation. One of the most important books for the whole movement, Pearl (2000), cites the Principle of the Common Cause as one of two sources of inspiration for the condition (which will be discussed below). Later on, it seemed to have become common knowledge that the PCC *follows from* the CMC; proofs of this fact are given e.g. by Williamson (2005) and Arntzenius (2005), while Eberhardt (2011) states that "Reichenbach's principle of common cause is a special case of the causal Markov condition when taken to apply to distributional properties". It is therefore quite surprising that one of the most distinguished writers on the subject, Clark Glymour, claims in a recent paper (Glymour (2010)) that "Neither, contrary to many commentators, does it [the CMC] imply Reichenbach's Principle of the Common Cause" (p. 175). Are all the proofs wrong, then? Or is the Principle they refer to something different from Reichenbach's PCC? The issue is important, since if the implication holds then any argument against the PCC is dangerous for the CMC, too. I will study this question in this chapter by providing the needed definitions (the presentation will be based mainly on Spirtes et al. (2000)), discussing the philosophical relationship between the CMC and the PCC, and presenting a proof regarding the CMC/PCC relationship. While the proof is based on Williamson (2005) (p. 52) I will present it in a different way, to highlight the fact that the main idea can be expressed without reference to any causal concepts.

---

1 Not because the interpretation of the probability is Bayesian, but because of the use of Bayes's theorem for updating probabilities.

I have to stress that this chapter is not meant to be a complete overview of the subjects of the Causal Markov Condition and causal structures. I will present the minimal amount of information needed to address the goal outlined in the previous paragraph; for a detailed introduction to these matters I would like to refer the Reader to the already mentioned Spirtes et al. (2000) and Pearl (2000).

It has to be mentioned that at least one influential approach to causation which involves probability, namely the interventionist theory of Woodward (2005), at the outset openly does not assume the CMC. However, in his book (p. 385) Woodward himself claims—with some reservations—that "it is far from obvious there are any real-life cases that violate" the CMC, and also aims to defend it from counterarguments by Cartwright (see the exchange in Hausman and Woodward (1999), Cartwright (2002) and Hausman and Woodward (2004), and also Steel (2006)).

## 4.1 DAGs—an Introduction

Let us start with the required definitions. A *directed graph* $G$ over $\mathbf{V}$ is a pair $\langle \mathbf{V}, \mathbf{E} \rangle$, where $\mathbf{V}$ is a set of nodes and $\mathbf{E}$ a set of arrows (ordered pairs of nodes). A *path* between nodes $X$ and $Y$ is a sequence of nodes beginning with $X$ and ending with $Y$ such that for any two nodes adjacent in the sequence there is an arrow between them (the direction does not matter). A node $X$ on a path is a *collider* if together with its adjacent nodes $Y$ and $Z$ in the path it forms an inverted fork: $Y \rightarrow X \leftarrow Z$. There is a *directed path* between vertices $X$ and $Y$, a fact symbolized by "$X \rightsquigarrow Y$", if there is an arrow between $X$ and $Y$ ("$X \rightarrow Y$") or there is some node $Z$ such that $X \rightsquigarrow Z$ and $Z \rightarrow Y$. A directed graph is *acyclic* if for any node $X$ it is not true that $X \rightsquigarrow X$. We will always assume that the nodes of any given graph represent random variables.

$Par(X)$, the set of "parents" of a node $X$, consists of the nodes $Z$ such that $Z \rightarrow X$. $Childr(X)$, the set of $X$'s "children", includes exactly the nodes $Z$ such that $X \rightarrow Z$. The sets $Anc(X)$ ("ancestors") and $Desc(X)$ ("descendants") are defined by substituting "$\rightsquigarrow$" for "$\rightarrow$" in the last two sentences—but with the addition that a node always is its own ancestor and descendant, but never its child or parent (see Spirtes et al. (2000), p. 10).

Not to stray from the recent literature, we will express the fact that variables $X$ and $Y$ are independent[2] as "$X \perp\!\!\!\perp Y$"; that they are independent given a third variable $Z$ as "$X \perp\!\!\!\perp Y \mid Z$"; and that they are *not* independent as "$X \not\perp\!\!\!\perp Y$".

The Markov Condition, in contrast to the Causal Markov Condition, is expressed exclusively by means of probabilistic and graph-related notions. It does not concern

2 See definition 5 in chapter 1, p. 1.

DAGs *per se*, but DAGs together with probability distributions over the set of their nodes.

**Definition 14 [Markov Condition]** A DAG $G$ over $\mathbf{V}$ and a probability distribution $P(\mathbf{V})$ *satisfy the Markov Condition* if and only if for any $W \in \mathbf{V}$,

$$W \perp\!\!\!\perp \mathbf{V} \setminus \left(Desc(W) \cup Par(W)\right) \mid Par(W).$$

In other words, in a graph which (together with a probability distribution over the set of its nodes) satisfies the Markov Condition (MC), every variable is independent of its nondescendants conditional on its parents. It is also independent (again, conditional on its parents) of its ancestors which are not its parents.

The MC can hold of DAGs and probability distributions with no appeal to any causality whatsoever. The Causal Markov Condition (CMC) holds of a subset of graph-probability distribution pairs for which MC holds; namely the graphs have to be "causal": they should represent a causal structure and the distribution to be "generated" by that structure. A *causal structure* for a population is a set of variables $\mathbf{V}$ together with a set $\mathbf{E}$ of ordered pairs of these variables, where a pair $\langle X, Y\rangle$ belongs to $\mathbf{E}$ whenever $X$ is a direct cause[3] of $Y$ relative to $\mathbf{V}$[4] (Spirtes et al. (2000), p. 22). Suppose we have a causal structure $C = \langle \mathbf{V}, \mathbf{E}\rangle$ and $P(\mathbf{V})$ is the actual probability distribution over $\mathbf{V}$; we then say that the distribution $P(\mathbf{V})$ *is generated* by the causal structure $C$. A causal structure $\langle \mathbf{V}, \mathbf{E}\rangle$ is *causally sufficient* for a given population iff it contains all common causes of any two variables in $\mathbf{V}$, apart from the ones which have the same value for all elements of the population. It is interesting that the definition of causal representation (Spirtes et al. (2000), p. 24) explicitly refers only to graphs representing causally sufficient causal structures; a DAG $G = \langle \mathbf{V}, \mathbf{E}\rangle$ *represents a causally sufficient causal structure* $C = \langle \mathbf{W}, \mathbf{F}\rangle$ if each node from $\mathbf{V}$ represents a variable from $\mathbf{W}$, every variable from $\mathbf{W}$ is represented by some node from $\mathbf{V}$, and there is an arrow between two vertices from $\mathbf{V}$ if and only if the ordered pair consisting of the two variables represented by the nodes at the rear end and the head of the arrow belongs to $\mathbf{F}$. A *causal graph* is then defined as a DAG which "represents a causal structure" (Spirtes et al. (2000), p. 24).[5]

3 It is interesting that the definition of a variable $X$ being a direct cause of variable $Y$ is in this framework a counterfactual definition; see Spirtes et al. (2000), p. 20.

4 This last qualification is important—if $X$ causes $Z$ by means of an intermediary variable $Y$, and yet we exclude $Y$ from our causal structure, then even though $X$ is not a direct cause of $Z$ "in general" we should have the pair $\langle X, Z\rangle$ in the set of pairs being the second element of our causal structure.

5 In fact, it does not seem that any significant generality is lost if a causal graph is thought to be coextensive with the structure it represents; in other words, if we can think of causal graphs simply *being* causal structures, and of its nodes *being* variables.

We take the quoted part of the definition of a causal graph to mean "a causally sufficient structure", since working on such structures seems to be the overall goal. The reasons are obvious; e.g., should one ignore a common cause $C$ for two correlated (but really directly causally unrelated) variables $A$ and $B$, one would be tempted to draw "$A \rightarrow B$" or "$B \rightarrow A$" in the causal graph, which would then give an incorrect picture of the real causal structure. Of course, it may be by no means evident what the "real" common causes are, and, *a fortiori*, which variables should be included for the causal structure to be sufficient. Nevertheless, we take causal graphs to be graphs representing causally sufficient structures; this decision will have no impact on the conclusions of this chapter.

## 4.2 The Causal Markov Condition

**Definition 15 [Causal Markov Condition]** A DAG $G$ over $\mathbf{V}$ and a probability distribution $P(\mathbf{V})$ *satisfy the Causal Markov Condition* if and only if
- $G$ is a causal graph;
- $P(\mathbf{V})$ is generated by the structure represented by $G$;
- $G$ and $P(\mathbf{V})$ satisfy the Markov Condition.

What is the relationship of the CMC and the Principle of the Common Cause, e.g. in the PCC 2 version, reformulated so that it would refer to random variables, and generalized so that the correlation would require the existence of a set of common causes?[6] Suppose there are only two variables in the causal structure, $X$ and $Y$. If they constitute a real-life counterexample to the PCC—that is, they are correlated, but there are no direct causal relations between them and there is in the world no set of variables which would render them conditionally independent—then the structure $C := \langle\{X, Y\}, \emptyset\rangle$ is causally sufficient for the given population. Therefore the graph $G := \langle\{X, Y\}, \emptyset\rangle$, with two vertices but no arrows[7], is of course a causal graph, but together with the real distribution over $X$ and $Y$, according to which $X$ and $Y$ are correlated, of course fails to meet the Markov Condition ($X$ is not independent of $Y$ conditional on the empty set) and so *a fortiori* the Causal Markov Condition.

In the other direction, suppose PCC 2 (in the reformulation hinted at above) is generally true, *and* that a causally sufficient structure is considered. The exogenous variables[8] have to be pairwise independent, since if they were not, then some of them

6 See Gyenis and Rédei (2010) for a rigorous translation of Reichenbach's ideas to the language of random variables; we will return to these matters at the end of chapter 5.

7 Again, the "$X$" in the graph is a node which represents the "$X$", a variable in the structure.

8 Meaning, the ones which are not pointed to by any arrow; the variables which have no causes in the structure considered. Such variables have to exist if the plausible assumption of finitude of the

would have to have (common) causes, and so some of them would not be exogenous after all. It is a well-known fact (for a proof see e.g. Steel (2005)) that the MC is true in any graph with independent exogenous variables. However, a stronger independence than pairwise independence is needed: due to the so called "Bernstein Paradox", there might be a dependence between sets of variables even if there is no dependence between any two variables[9]. Therefore, to arrive at an implication from the PCC to the CMC, one would have to reformulate the principle not only so that it considered random variables and more than one common cause, but also generalized the starting point from the correlation of two variables to the existence of a correlated set of variables (with no direct causal relationships).

We would like to show that some arguments offered in the literature as proving some version of the PCC on the basis of CMC can be expressed without reference to any causal notions. For example, the following principle is called by Williamson (2005) a "Principle of the Common Cause".

**Definition 16 [Principle of the Common Ancestor]** The *Principle of the Common Ancestor* holds of a DAG $G$ over $\mathbf{V}$ and a probability distribution $P(\mathbf{V})$ if, whenever $A \rightleftharpoons B$, then $A \rightsquigarrow B$ or $B \rightsquigarrow A$ or there is a $U \subseteq \mathbf{V}$ such that $C \in U$ implies $C \rightsquigarrow A$ and $C \rightsquigarrow B$, and $A \perp\!\!\!\perp B \mid U$.

The shortest (known to us) proof of the relationship between the Markov Condition and the Principle of the Common Ancestor uses the notion of *d*-separation (Pearl (1988); we use the definition from Spirtes et al. (2000), p. 14). (The "*d*" is from "directional"; the notion of *d*-separation is highly technical and not easy to illustrate intuitively—see e.g. chapter 3.7.1 of Spirtes et al. (2000).) Consider a graph $G$. If $X$ and $Y$ are distinct vertices of $G$ and $\mathbf{W}$ is a set of vertices of $G$ containing neither $X$ nor $Y$, then $X$ and $Y$ are d-*separated given* $\mathbf{W}$ ($\mathbf{W}$ d-*separates* $X$ and $Y$) in $G$ if and only if there exists no path $U$ between $X$ and $Y$ such that

- every collider on $U$ has a descendant in $\mathbf{W}$;
- no other vertex on $U$ belongs to $\mathbf{W}$.

The notion of *d*-separation is very useful when discussing Bayesian networks due e.g. to the following fact, which we will need later on (we cite here the formulation from Williamson (2005), p. 17):

causal structure (we can only measure a finite number of variables and adding an infinite number of unmeasured variables to the given structure would require some serious argument) is made, due to the fact that the graphs are to be acyclic.

9 For a discussion of this paradox in the context of common causes, see Uffink (1999).

**Fact 8** (Verma and Pearl (1988)). *Given a DAG $G$ over $\mathbf{V}$ and $R, S, T \subseteq \mathbf{V}$, $T$ d-separates $R$ and $S$ if and only if $R \perp\!\!\!\perp S \mid T$ for all probability distributions $P(\mathbf{V})$ which together with $G$ satisfy the Markov Condition.*

For example, if we know that a DAG with a probability distribution satisfies the Markov condition, and we find that distinct variables $X$ and $Y$ are $d$-separated by $\emptyset$, we can infer that $X$ and $Y$ are not correlated.

The following fact is a direct companion to Proposition 4.1 from Williamson (2005), p. 52; it has (together with the proof) been only reworded so that it does not refer to causal notions.

**Fact 9.** *The Markov Condition implies the Principle of the Common Ancestor.*

*Proof.* Suppose the Markov Condition holds for a graph $G$ over $\mathbf{V}$ and a probability distribution $P(\mathbf{V})$. Let $A, B \in \mathbf{V}$. Suppose it is not the case that ($A \rightsquigarrow B$ or $B \rightsquigarrow A$ or there is a $C \in \mathbf{V}$ such that $C \rightsquigarrow A$ and $C \rightsquigarrow B$). Then variables $A$ and $B$ are $d$-separated by $\emptyset$, since any path between them has to include a collider. In such a case, $A \perp\!\!\!\perp B$.

Suppose, then, that $A \rightleftharpoons B$. From the last paragraph we infer by contraposition that, if it is not the case that $A \rightsquigarrow B$ or $B \rightsquigarrow A$ (when the Principle would be trivially true), there has to be at least one $C \in \mathbf{V}$ such that $C \rightsquigarrow A$ and $C \rightsquigarrow B$. Let $\mathbf{U}$ be the set of all such $C$s. $\mathbf{U}$ $d$-separates $A$ and $B$, so $A \perp\!\!\!\perp B \mid \mathbf{U}$. □

Of course, if the graph under consideration is causal, the Principle of the Common Ancestor becomes a version of the PCC, claiming the existence of a set of common causes for the correlated variables, which act as screeners for those variables.

## 4.3 Conclusions

We have seen that real-life counterexamples to the PCC would lead to failure of the CMC. We have also presented a general proof that in any graph which satisfies the MC a certain principle is valid; if the graph is a causal graph (and the distribution is the one generated by the structure) then this principle becomes a version of the PCC. What is, then, the reason for the already mentioned claim of Glymour (2010) that Reichenbach's PCC does not follow from the CMC?

Perhaps the matter is simple and the word "Reichenbach" is the key; notice the complete absence of the statistical relevance conditions from the considerations of these chapter. This is of course reasonable; if we speak about correlated events, we can consider a cause raising the probability of the events; but the correlation of two variables (e.g. $X$ and $Y$) typically leads to numerous correlations between events (e.g. "$X = 1$" and "$Y = 1$" and so on, for various (but maybe not all) values of the variables). But let us consider two correlated *binary* variables[10], which do not influence each other directly. One can look at the existence of a common cause variable for such two variables as the existence of a common screener-off for the correlations between the appropriate events. But, as we know from example 2 in section 2.2, it might very well be that no element of the screener-off is positively statistically relevant for any of the correlated events, so Reichenbach's conditions (even as generalized as in the definition of a statistical common cause system) cannot be satisfied.

**10** Recall the close correspondence between binary variables and events, section 1.1, p. 1 above.

# 5 Causal Closedness

(The material in sections 5.2-5.6 originates from joint research carried in 2008-2009 by Michał Marczyk and the author, published in Wroński and Marczyk (2013).)

## 5.1 A Preliminary Formal Remark

For most of the results of this chapter, the sample spaces of the probability spaces involved are irrelevant. The crucial factors are the Boolean algebra being the event space and the measure defined on that algebra. Therefore—until section 5.10—if no other qualification is given, a probability space is meant to be a pair $\langle S, P \rangle$, where $S$ is a Boolean algebra[1] and $P$ is a classical measure on $S$. In section 5.6 nonclassical spaces are considered, in which the Boolean algebra is exchanged for a nondistributive orthomodular lattice (OML). The required definitions are presented. By $\mathbf{1}_S$ we refer to the unity of $S$, whether it is a Boolean Algebra or a nondistributive OML.

Also, throughout this chapter, by a "common cause" we always mean a "statistical common cause". At the beginning we usually supply the additional adjective, but then sometimes refrain from using it to conserve space, as the arguments unfortunately become rather cluttered even without the additional vocabulary.

## 5.2 Causal (up-to-$n$-)closedness

### 5.2.1 Introduction

Suppose a probability space contains a correlation between two events we believe to be causally independent. Does the space contain a common cause for the correlation? If not, can the probability space be extended to contain such a cause but 'preserving' the old measure? This question has been asked and positively answered in Hofer-Szabó et al. (1999), where the notion of *common cause completability* was introduced. Speaking a bit informally, a probability space $S$ is said to be common cause completable with respect to a set $\mathbf{A}$ of pairs of correlated events iff there exists an extension of the space containing statistical common causes of all the correlated pairs in $\mathbf{A}$. Gyenis and Rédei (2004) introduced the notion of *common cause closedness*, which (in our slightly different terminology) is equivalent to the following: a probability space $S$ is common cause closed (or "causally closed") with respect to a relation of independence $R_{ind} \subseteq S^2$ iff it contains statistical common causes (recall definition 9, p. 30)

1 We omit the usual requirement of $\sigma$-completeness because, while the notion of causal up-to-$n$-closedness will be general, the results proved regarding it will concern the finite cases only.

for all pairs of correlated events belonging to $R_{ind}$. The authors have proven therein that a finite classical probability space with no atoms of probability 0 is non-trivially common cause closed w.r.t. the relation of logical independence iff it is the space consisting of a Boolean algebra with 5 atoms and the uniform probability measure.[2] In other words, finite classical probability spaces (big enough to contain correlations between logically independent events) are in general *not* common cause closed w.r.t. the relation of logical independence, i.e. they contain a correlation between logically independent events for which no statistical common cause in the space exists; the only exception to this rule is the space with precisely 5 atoms of probability $\frac{1}{5}$ each. More spaces are common cause closed w.r.t. a more stringent relation of logical independence modulo measure zero events ("$L^{+}_{ind}$", see definition 18 below): they are the spaces with 5 atoms of probability $\frac{1}{5}$ each and any number of atoms of probability 0.

Still, a (statistical) common cause is not the only entity which could be used as an explanation for a correlation. As we mentioned earlier in chapter 2, Hofer-Szabó and Rédei (2004) generalized the idea of a statistical common cause, arriving at *statistical common cause systems* ("SCCSs"; recall definition 10, p. 30). As already noted, SCCSs may have any countable size greater than 1; the special case of size 2 reduces to the usual notion of common cause.

It was natural for corresponding notions of causal closedness to be introduced; a probability space is said to be *causally n-closed*[3] w.r.t. a relation of independence $R_{ind}$ iff it contains an SCCS of size $n$ for any correlation between $A, B$ such that $\langle A, B\rangle \in R_{ind}$. It is one of the results of the present chapter that with the exception of the 5-atom uniform distribution probability space, no finite probability spaces without 0-probability atoms are causally $n$-closed w.r.t. the relation of logical independence, for any $n \geqslant 2$. Similarly, with the exception of the spaces with 5 atoms of probability $\frac{1}{5}$ each and any number of atoms of probability 0, no finite probability spaces with 0-probability atoms are causally $n$-closed w.r.t. $L^{+}_{ind}$, for any $n \geqslant 2$.

We are interested in a slightly different version of causal closedness. If the overarching goal is to find explanations for correlations, why should we expect all explanations to be SCCSs of the same size? Perhaps some correlations are explained by common causes and others by SCCSs of a bigger size. We propose to explore the idea of *causal up-to-n-closedness*—a probability space is causally up-to-$n$-closed w.r.t. a relation of independence $R_{ind}$ iff it contains an SCCS of size at most $n$ for any correlation between events $A, B$ such that $\langle A, B\rangle \in R_{ind}$.

---

2 The phrasing of the paper was in fact stronger, omitting the assumption about non-0 probabilities on the atoms (due to a missed special sub-case in the proof of case 3 of proposition 4 on p. 1299). The issue is connected to the distinction between proper and improper common causes and is discussed below in section 5.4.

3 The notion was introduced in Hofer-Szabó and Rédei (2006).

It turns out that, in the class of finite classical probability spaces with no atoms of probability 0, just as the space with 5 atoms and the uniform measure is unique with regard to common cause closedness, the whole class of spaces with uniform distribution is special with regard to causal up-to-3-closedness—see theorem 4: a finite classical probability space with no atoms of probability 0 has the uniform distribution iff it is causally up-to-3-closed w.r.t. the relation of logical independence. We provide a method of constructing a statistical common cause or an SCCS of size 3 for any correlation between logically independent events in any finite classical probability space with the uniform distribution.

We require (following Gyenis and Rédei) of a causally closed probability space that all correlations be explained by means of proper—that is, differing from both correlated events by a non-zero measure event—statistical common causes. This results in the fact that a space causally closed w.r.t. the relation of logical independence can be transformed into a space which is not causally closed w.r.t. this relation, just by adding a 0-probability atom. Perhaps, to avoid this unfortunate consequence, the notion of logical independence modulo measure zero events should be required? We discuss the matter in section 5.4.

In this chapter we also briefly consider other independence relations, a generalization of our results to finite non-classical probability spaces, and closedness w.r.t. to the more general *deductive explanantes*. Lastly, we briefly report some known results on causal closedness of atomless spaces we will use in the next chapter.

### 5.2.2 Preliminary Definitions

In the following assume that we are given a finite classical probability space $\langle S, P\rangle$, where $S$ is a finite Boolean algebra and $P$ is a classical measure on $S$. By Stone's representation theorem, $S$ is isomorphic—and may be identified with—the algebra of all subsets of the set $\{0, \dots, n-1\}$ for some $n \in \mathbb{N}$.

In the sequel we will sometimes consider spaces of the form $\langle S^+, P^+\rangle$, where $S^+$ and $P^+$ are as defined below:

**Definition 17** Let $\langle S, P\rangle$ be a finite classical probability space. $S^+$ is the unique Boolean algebra whose set of atoms consists of all the non-zero probability atoms of $S$. $P^+$ is the restriction of $P$ to $S^+$.

We will now define two relations of logical independence. Intuitively, we will regard two events as logically independent if when we learn that one of the events occurs (or does not occur) we cannot infer that the other occurs (or does not occur), for all four Boolean combinations.

**Definition 18 [Logical independence]** We say that events $A, B \in S$ are *logically independent* ($\langle A, B\rangle \in L_{ind}$) iff all of the following sets are nonempty:

- $A \cap B$;
- $A \cap B^{\perp}$;
- $A^{\perp} \cap B$;
- $A^{\perp} \cap B^{\perp}$.

We say that events $A, B \in S$ are *logically independent modulo measure zero events* ($\langle A, B \rangle \in L^{+}_{ind}$) iff all of the following numbers are positive:

- $P(A \cap B)$;
- $P(A \cap B^{\perp})$;
- $P(A^{\perp} \cap B)$;
- $P(A^{\perp} \cap B^{\perp})$.

Equivalently, two events are logically independent if neither of the events is contained in the other one, their intersection is nonempty and the sum of the two is less than the whole space. Two events are logically independent modulo measure zero events if every Boolean combination of them has a non-zero probability of occurring. It is always true that $L^{+}_{ind} \subseteq L_{ind}$; if there are 0-probability atoms in the space, the inclusion may be strict.

Since both of the discussed relations are symmetric, we will frequently omit the chevrons "$\langle$" and "$\rangle$" when referring to pairs of independent events.

The following definition is a refinement of the SCC idea, expressing the requirement that a common cause should be meaningfully different from both correlated events.

**Definition 19 [Proper SCC(S)]** A statistical common cause $C$ of events $A$ and $B$ is a *proper* statistical common cause of $A$ and $B$ if it differs from both $A$ and $B$ by more than a measure zero event. It is an *improper* SCC of these events otherwise.

An SCCS $\{C_i\}_{i \in I}$ of events $A$ and $B$ is a *proper* SCCS of $A$ and $B$ if all its elements differ from both $A$ and $B$ by more than a measure zero event. It is an *improper* SCCS of these events otherwise.

We will sometimes say that a probability space *contains* an SCCS, which means that the SCCS is a partition of unity of the underlying algebra of the space.

We now come to the main topic of this chapter. Should someone prefer it, the following definition could be phrased in terms of SCCSs only.

**Definition 20 [Causal up-to-$n$-closedness]** We say that a classical probability space is *causally up-to-n-closed* w.r.t. to a relation of independence $R_{ind}$ if all pairs of correlated events independent in the sense of $R_{ind}$ possess a proper statistical common cause or a proper statistical common cause system of size at most $n$.

If the space is causally up-to-2-closed, we also say that it is *causally closed* or *common cause closed*.

### 5.2.3 Summary of the Results of this Chapter

| | $\langle S, P\rangle$ is up-to-3-closed w.r.t. | |
|---|---|---|
| | $L_{ind}$ | $L^{+}_{ind}$ |
| $P$ is uniform | $\Rightarrow$ (10) | $\Rightarrow$ (10) |
| $P^{+}$ is uniform | $\Leftarrow$ (11)<br>$\Rightarrow^{*}$ (12) | $\Leftrightarrow$ (11,12) |

**Table 5.1:** The main results of the chapter. The numbers in parentheses correspond to lemmas below. The star means that the implication does not hold in full generality: some pairs of correlated pairs which differ only by a measure zero event have only improper SCCs.

Theorem 3 will be our main tool in proving the lemmas featured in table 5.1.

**Theorem 3.** *Let $\langle S, P\rangle$ be a finite classical probability space with $S^{+}$ having at least 4 atoms of non-zero probability. Then $P^{+}$ is uniform if and only if $\langle S^{+}, P^{+}\rangle$ is causally up-to-3-closed w.r.t. $L^{+}_{ind}$.*

Lemmas 10-12 tie uniformity of $P$ and $P^{+}$ with causal up-to-3-closedness of $\langle S, P\rangle$ with respect to the two notions of independence introduced above.

**Lemma 10.** *Let $\langle S, P\rangle$ be a finite classical probability space with $S$ having at least 4 atoms. If $P$ is uniform, then $\langle S, P\rangle$ is causally up-to-3-closed w.r.t. $L_{ind}$ and $L^{+}_{ind}$.*

**Lemma 11.** *Let $\langle S, P\rangle$ be a finite classical probability space with $S^{+}$ having at least 4 atoms. If $P^{+}$ is not uniform, then $\langle S, P\rangle$ is not causally up-to-3-closed w.r.t. either $L_{ind}$ or $L^{+}_{ind}$.*

**Lemma 12.** *Let $\langle S, P\rangle$ be a finite classical probability space with $S^{+}$ having at least 4 atoms. If $P^{+}$ is uniform, then $\langle S, P\rangle$ is causally up-to-3-closed w.r.t. $L^{+}_{ind}$. All correlated pairs from $L_{ind} \setminus L^{+}_{ind}$ have statistical common causes, but some only have improper ones.*

## 5.3 Proofs

### 5.3.1 Some Useful Parameters

For expository reasons, we will not prove theorem 3 directly but rather show its equivalent, theorem 4 (p. 75). Before proceeding with the proof, we shall introduce a few useful parameters one may associate with a pair of events $A$, $B$ in a *finite* classical probability space $\langle S, P\rangle$.

Let $n$ be the number of atoms in the Boolean algebra $S$. The size of the set of atoms lying below $A$ in the lattice ordering of $S$ will from now on be referred to as $a$, and likewise for $B$ and $b$. The analogous parameter associated with the conjunction of events $A$ and $B$ is just the size of the intersection of the relevant sets of atoms and will be called $k$.

It will soon become apparent that while $a$ and $b$ have some utility in the discussion to follow, the more convenient parameters describe $A$ and $B$ in terms of the number of atoms belonging to one but not the other. Thus we let $a' = a - k$ and $b' = b - k$. In fact, if we set $z = n - (a' + k + b')$, we obtain a set of four numbers precisely describing the blocks of the partition of the set of atoms of $S$ into the four classes which need to be nonempty for $A$ and $B$ to be logically independent. It is clear that in the case of logically independent events $a'$, $b'$, $k$ and $z$ are all non-zero.

Lastly, before we begin the proof of the main result of this chapter, let us recall corollary 4, p. 34: when searching for statistical common causes, screening off is enough. If both an event and its complement screen off a correlation, then one of them is a statistical common cause for the correlation.

### 5.3.2 Proof of Theorem 3

In this section we will provide a proof of the main tool in this chapter—theorem 3, formulated in section 5.2.3. The form in which it was stated in that section is dictated by its use in the proofs of lemmas 10-12. However, when treated in isolation, it is better versed in the following way:

**Theorem 4** (Wroński and Marczyk (2013), equivalent to theorem 3). *Let $\langle S, P\rangle$ be a finite classical probability space with no atoms of probability 0. Suppose $S$ has at least 4 atoms.*[4] *The following conditions are equivalent:*

***Measure uniformity:*** *$P$ is the uniform probability measure on $S$;*

***Causal up-to-3-closedness w.r.t. $L_{ind}$:*** *$\langle S, P\rangle$ is causally up-to-3-closed w.r.t. the relation of logical independence.*

Before proceeding with the proof we will provide a sketch of the construction and some requisite definitions. Instead of focusing on a particular $n$-atom algebra, we will show how the problem presents itself while we "move" from smaller to bigger algebras. We assume without loss of generality that the set of atoms of an $n$-atom Boolean algebra

---

4 It is easy to verify that if $S$ has 3 atoms or less, then $\langle S, P\rangle$ contains no correlations between logically independent events.

is $\{0, 1, \cdots, n-1\}$ and that each event is a set of atoms. Consider the sequence of all finite classical probability spaces with the uniform probability measure, in which the number of atoms of the underlying Boolean algebra of the space increases by 1 at each step, beginning with the algebra with a single atom. We use the shorthand expression "at stage $n$" to mean "in the probability space with uniform distribution whose underlying Boolean algebra has $n$ atoms". Observe that due to our convention whereby events are identified with sets of atoms, an event present at stage $m$ (one found in the algebra from that stage) is also present at all further stages. In other words, a set of atoms defining an event at stage $m$ can also be interpreted as defining an event at any stage $m'$, with $m' > m$. Thus we can naturally say that a certain event belongs to many different probability spaces; e.g. the event $\{1, 2, 11\}$ is present at stages 12, 13, and so on. Similarly, pairs of events can be present at many stages—and be correlated at some, but not at others. If they are correlated at stage $m$, they are correlated at all stages $n$, for $n > m$ (see below). A similar thing is true of logical independence: a pair may not consist of logically independent events at stage $n$, because their union is the whole set of $n$ atoms, but may become a pair of logically independent events at stage $n+1$, when an additional atom is introduced which does not belong to either of the events in question.

Some remarks on the shape of events considered are in order. We will always be talking about pairs of events $A, B$, with numbers $a$, $a'$, $b$, $b'$, $k$, $z$ and $n$ defined as above (see section 5.3.1). We assume $a \geqslant b$. Also, since we are dealing with the uniform measure, all relevant characteristics of a pair of events $A, B$ are determined by the numbers $a'$, $b'$, $k$, and $z$; therefore, for any combination of these numbers it is sufficient only to consider a single example of a pair displaying them. The rest is just a matter of renaming the atoms. For example, if we are looking for an explanation for the pair $\{\{8, 7, 3, 5\}, \{2, 8, 7\}\}$ at stage 10, or the pair $\{\{1, 3, 5, 6\}, \{1, 6, 4\}\}$ at the same stage, we shall search for an explanation for the pair $\{\{0, 1, 2, 3\}, \{2, 3, 4\}\}$ at stage 10 and then just appropriately "translate" the result (explicit examples of this follow in section 5.3.2.1). In general: the convention we adopt is for $A$ to be a set of consecutive atoms beginning with 0, and $B$ a set of consecutive atoms beginning with $a-k$.

For illustrative purposes we propose to examine the situation at the early stages. The proof proper begins with definition 21 below. For the remainder of section 5.3.2, by "common cause" we will always mean "proper common cause"; similarly with "common cause system".

There are no correlated pairs of logically independent events at stage 1; similarly for stages 2, 3 and 4. (Remember the measure is uniform and so at stage 4 e.g. the pair $\{\{0, 1\}, \{1, 2\}\}$, while composed of logically independent events, is not correlated.)

The first correlated pairs of logically independent events appear at stage 5. These are of one of the two following types: either $a' = b' = k = 1$, or $a' = b' = 1$ and $k = 2$. Proposition 3 from Gyenis and Rédei (2004) says that all pairs of these types have statistical common causes at stage 5. As noted above, we can without loss of generality consider just two tokens of these types—the pairs $\{\{0,1\},\{1,2\}\}$ and $\{\{0,1,2\},\{1,2,3\}\}$. In the first case, the events already formed a logically independent pair at stage 4 but were not correlated—we will say that the pair *appears from below at stage* 5 (see definition 21 below). In the second case, stage 5 is the first stage where the events form a logically independent pair and they are already correlated at that stage. We will say that the pair $\{\{0,1,2\},\{1,2,3\}\}$ *appears from above at stage* 5. There are no other correlated pairs of logically independent events at stage 5. It will turn out that we can always find statistical common causes for pairs which appear from above or from below at a given stage.

Let us move to stage 6. A new (type of) pair appears from above—$\{\{0,1,2,3\},\{1,2,3,4\}\}$. No pairs appear from below but both pairs which appeared at stage 5 are still correlated and logically independent at stage 6 (as well as at all later stages), so they are again in need of an explanation at this higher stage. It turns out that if a correlated pair of logically independent events at stage $n$ is "inherited" from the earlier stages, i.e. it appears neither from above nor from below at stage $n$, we can modify the common cause which we know how to supply for it at the stage where it originally appeared to provide it with an explanation adequate at stage $n$. This takes the form of a statistical common cause or, in some cases, an SCCS of size 3.

**Definition 21 [Appearing from above or below]** A pair $\{A,B\}$ of events *appears from above at stage n* if it is (1) logically independent at stage $n$, (2) not logically independent at stage $n-1$ and (3) correlated at stage $n$.

A pair $\{A,B\}$ of events *appears from below at stage n* if it is (1) logically independent at stage $n$, (2) logically independent at stage $n-1$ and (3) correlated at stage $n$, but (4) not correlated at stage $n-1$.

We will divide common causes into types depending on whether the occurrence of a given common cause makes the occurrence of at least one member of the correlation it explains necessary, impossible, or possible with probability less then 1.[5]

**Definition 22 [1-, 0-, and #-type statistical common causes]** A proper statistical common cause $C$ for a correlated pair of logically independent events $A, B$ is said to be:

- *1-type* iff $P(A \mid C) = 1$ or $P(B \mid C) = 1$;
- *0-type* iff $P(A \mid C^{\perp}) = 0$ or $P(B \mid C^{\perp}) = 0$;
- *#-type* iff it is neither 1-type nor 0-type.

5 Since the context of theorem 4 is that of finite spaces, the difference between necessity and probability 1 can be dismissed.

Notice that no statistical common cause $C$ for some two logically independent, correlated events $A$ and $B$ can be both 1-type and 0-type at the same time.

**Definition 23 [0-type statistical common cause system]** A proper statistical common cause system of size $n$ $\{C_i\}_{i \in \{0,\dots,n-1\}}$ is a *0-type statistical common cause system* (*0-type SCCS*) for the correlation iff $P(A \mid C_{n-1}) = 0$ or $P(B \mid C_{n-1}) = 0$.

We do not need to worry about the fact that rearranging the elements of a 0-type SCCS necessarily makes it lose the 0-type status, because during the proof the SCCSs will be explicitly construed so that their "last" element gives conditional probability 0 to both correlated events to be explained. Were this notion to be used in general its definition should be rephrased as an existential condition: "there exists $m \leqslant n - 1$ such that $P(A \mid C_m) = 0$ and $P(B \mid C_m) = 0$".

We will prove the following:

- if a pair appears from above at stage $n$, it has a statistical common cause at that stage (lemma 14);
- if a pair appears from below at stage $n$, it has a statistical common cause at that stage (lemma 15);
- if a pair of logically independent events is correlated at stage $n$ and has a statistical common cause or a 0-type SCCS of size 3 at that stage, it has a statistical common cause or a 0-type SCCS of size 3 at stage $n + 1$ (lemma 16).

It should be straightforward to see that this is enough to prove theorem 4 (p. 75) in its "downward" direction. Consider a correlated pair of logically independent events $A, B$ at stage $n$. If it appears from above we produce a common cause using the technique described in lemma 14. If it appears from below we use the method from lemma 15. If it appears neither from above nor from below it means that it was logically independent at stage $n-1$ and was correlated at that stage, and we repeat the question at stage $n-1$. This descent terminates at the stage where our pair first appeared, which clearly must have been either from below or from above. This allows us to apply either lemma 14 or lemma 15, as appropriate, followed by lemma 16 to move back up to stage $n$, where we will now be able to supply the pair with an SCC or an SCCS of size 3. As said before, the SCCs and SCCSs we will construct will always be *proper* SCCs and SCCSs.

Put $Corr(A, B) := P(AB) - P(A)P(B)$. $Corr(A, B)$ can always be expressed as a fraction with the denominator being $n^2$. Of special interest to us will be the numerator of this fraction. Let us call this number $SC_n(A, B)$. (For example, if $A = \{0, 1, 2\}$ and $B = \{2, 3\}$, $SC_5(A, B) = -1$.) If $SC_n(A, B) \leqslant 0$, the events are not correlated at stage $n$. If $SC_n(A, B) > 0$, $A$ and $B$ are correlated at stage $n$ and we need to find either a common cause or a common cause system of size 3 for them. The following lemma will aid us in our endeavour (remember the definitions from section 5.3.1):

**Lemma 13.** *Let $\langle S_n, P\rangle$ be a finite classical probability space, $S_n$ being the Boolean algebra with n atoms and P the uniform measure on $S_n$. Let $A, B \in S_n$. Then $SC_n(A, B) = kz - a'b'$.*

*Proof.* $Corr(A, B) = P(AB) - P(A)P(B) = \frac{k}{n} - \frac{k+a'}{n}\frac{k+b'}{n} = \frac{k(n-k-a'-b')-a'b'}{n^2} = \frac{kz-a'b'}{n^2}$. Therefore $SC_n(A, B) = kz - a'b'$. □

An immediate consequence of this lemma is that any pair of logically independent events will eventually (at a high enough stage) be correlated – it is just a matter of injecting enough atoms into $z$. For example, consider events $A = \{0, 1, 2, 3, 4, 5, 6\}$, $B = \{6, 7, 8, 9, 10, 11\}$. At any stage $n$, $SC_n(A, B)$ is equal to $z - 30$. This means that the pair is correlated at all stages in which $z > 30$; in other words, at stages 43 and up. At some earlier stages (from 13 to 42) the pair is logically independent but not correlated; at stage 12 it is not logically independent; finally, the events constituting it do not fit in the algebras from stages lower than that.

Notice that since for any $A, B$: $SC_{n+1}(A, B) = SC_n(A, B) + k$, it follows that at the stage $m$ where the pair first appears (either from above or from below) $SC_m(A, B)$ is positive but less than or equal to $k$.

We now have all tools we need to prove theorem 4.

*Proof.* (of theorem 4)

**Measure uniformity ⇒ Causal up-to-3-closedness w.r.t.** $L_{ind}$

**Lemma 14.** *Suppose a pair A, B appears from above at stage n. Then there exists a 1-type common cause for the correlation at that stage.*

*Proof.* We are at stage $n$. Since the pair $A, B$ appears from above at this stage, $z = 1$ and so (by lemma 13) $SC_n(A, B) = k - a'b'$. (If $z$ was equal to 0 the events would not be logically independent at stage $n$; if it was greater than 1 the events would be logically independent at stage $n-1$ too, and so the pair would not appear from above at stage $n$.) Notice that since $A, B$ are logically independent (so both $a'$ and $b'$ are non-zero) but correlated at stage $n$, $0 < SC_n(A, B) = k - a'b' < k$. Let $C$ consist of exactly $SC_n(A, B)$ atoms from the intersection $A \cap B$. Such a $C$ will be a screener-off for the correlation, since $P(AB \mid C) = 1 = P(A \mid C)P(B \mid C)$. What remains is to show that $C^\perp$ is a screener-off as well. This follows from the observation that $P(AB \mid C^\perp) = \frac{k-(k-a'b')}{n-(k-a'b')} = \frac{a'b'}{n-k+a'b'} = \frac{a'b'(n-k+a'b')}{(n-k+a'b')^2} = \frac{a'b'(1+a'+b'+k)-a'b'k+a'^2b'^2}{(n-k+a'b')^2} = \frac{a'b'+a'b'^2+a'^2b'+a'^2b'^2}{(n-k+a'b')^2} = \frac{a'+a'b'}{n-k+a'b'} \cdot \frac{b'+a'b'}{n-k+a'b'} = \frac{k+a'-(k-a'b')}{n-k+a'b'} \cdot \frac{k+b'-(k-a'b')}{n-k+a'b'} = \frac{k+a'-SC_n(A,B)}{n-k+a'b'} \cdot \frac{k+b'-SC_n(A,B)}{n-k+a'b'} = P(A \mid C^\perp)P(B \mid C^\perp)$. □

**Lemma 15.** *Suppose a pair A, B appears from below at stage n. Then there exists a 1-type common cause or a 0-type common cause for the correlation at that stage.*

*Proof.* **Case 1: $k > b'$ and $a' > z$.** In this case we will construct a 1-type common cause. Let $C$ consist of $k - b'$ atoms from $A \cap B$ and $a' - z$ atoms from $A \setminus B$. Since $C \subset A$, it screens off the correlation: $P(AB \mid C) = P(B \mid C) = 1 \cdot P(B \mid C) = P(A \mid C)P(B \mid C)$. We need to show that $C^\perp$ screens off the correlation as well. This follows from the fact that $P(AB \mid C^{\perp}) = \frac{b'}{n-(k-b')-(a'-z)} = \frac{b'}{2b'+2z} = \frac{2b'^2+2zb'}{(2b'+2z)^2} = \frac{(b'+z)2b'}{(2b'+2z)^2} = \frac{b'+z}{2b'+2z} \cdot \frac{2b'}{2b'+2z} = \frac{b'+z}{n-(k-b')-(a'-z)} \cdot \frac{2b'}{n-(k-b')-(a'-z)} = P(A \mid C^{\perp})P(B \mid C^{\perp})$.

**Case 2: $z > b'$ and $a' > k$.** In this case we will construct a 0-type common cause. Let $C^{\perp}$ consist of $a'-k$ atoms from $A \setminus B$ and $z-b'$ atoms from $(A \cup B)^{\perp}$. Since $C^{\perp} \subset B^{\perp}$, it screens off the correlation: $P(AB \mid C^{\perp}) = 0 = P(A \mid C^{\perp}) \cdot 0 = P(A \mid C^{\perp})P(B \mid C^{\perp})$. We need to show that $C$ too screens off the correlation. This follows from the fact that $P(AB \mid C) = \frac{k}{n-(a'-k)-(z-b')} = \frac{k}{2k+2b'} = \frac{2k^2+2kb'}{(2k+2b')^2} = \frac{2k(k+b')}{(2k+2b')^2} = \frac{2k}{2k+2b'} \cdot \frac{k+b'}{2k+2b'} = \frac{2k}{n-(a'-k)-(z-b')} \cdot \frac{k+b'}{n-(a'-k)-(z-b')} = P(A \mid C)P(B \mid C)$.

**Case 3a: $z \geqslant a'$, $k \geqslant a'$ and $a' > b'$.**

As can be verified easily, in this case $k = z = a'$ and $b' = a' - 1$. We can construct both a 0-type common cause and a 1-type common cause. Suppose we choose to produce the former. An appropriate $C^{\perp}$ would consist of just a single atom from $(A \cup B)^{\perp}$; $C^{\perp}$ screens off the correlation because $P(AB \mid C^{\perp}) = 0 = P(A \mid C^{\perp})P(B \mid C^{\perp})$. That $C$ is also a screener-off is guaranteed by the fact that $P(AB \mid C) - P(A \mid C)P(B \mid C) = \frac{k}{k+a'+b'+z-1} - \frac{k+a'}{k+a'+b'+z-1} \cdot \frac{k+b'}{k+a'+b'+z-1} = \frac{k}{4k-2} - \frac{2k}{2(2k-1)} \cdot \frac{2k-1}{4k-2} = 0$.

To produce a 1-type common cause instead, let $C$ consist of just a single atom from $A \cap B$; $C$ screens off the correlation because $P(AB \mid C) = 1 = P(A \mid C)P(B \mid C)$. That $C^{\perp}$ is also a screener-off follows from the fact that $P(AB \mid C^{\perp}) = \frac{k-1}{k-1+a'+b'+z} = \frac{b'}{2b'+2a'} = \frac{2b'^2+2a'b'}{(2b'+2a')^2} = \frac{(a'+b')2b'}{(2b'+2a')^2} = \frac{a'+b'}{2b'+2a'} \cdot \frac{2b'}{2b'+2a'} = \frac{k-1+a'}{2b'+2a'} \cdot \frac{k-1+b'}{2b'+2a'} = P(A \mid C^{\perp})P(B \mid C^{\perp})$.

**Case 3b: $z = a' + 1$ and $k = a' = b'$.**

In this case we will construct a 0-type common cause. Let $C^{\perp}$ consist of just a single atom from $(A \cup B)^{\perp}$; $C^{\perp}$ screens off the correlation because $P(AB \mid C^{\perp}) = 0 = P(A \mid C^{\perp})P(B \mid C^{\perp})$. $C$ screens off the correlation because $P(AB \mid C) = \frac{k}{4k} = \frac{4k^2}{16k^2} = \frac{2k}{4k} \cdot \frac{2k}{4k} = \frac{k+a'}{k+a'+b'+z-1} \cdot \frac{k+b'}{k+a'+b'+z-1} = P(A \mid C)P(B \mid C)$.

**Case 3c: $k = a' + 1$ and $z = a' = b'$.**

In this case we will construct a 1-type common cause. Let $C$ consist of just a single atom from $A \cap B$; as in case 3a, $C$ screens off the correlation. That $C^{\perp}$ is also a screener-off follows from $P(AB \mid C^{\perp}) = \frac{a'}{4a'} = \frac{4a'^2}{16a'^2} = \frac{2a'}{4a'} \cdot \frac{2a'}{4a'} = \frac{k-1+a'}{k-1+a'+b'+z} \cdot \frac{k-1+b'}{k-1+a'+b'+z} = P(A \mid C^{\perp})P(B \mid C^{\perp})$. □

Notice that the five cases used in the proof above are exhaustive. For example (due to lemma 13), if $k = a'$, then $z = b' + 1$. (Were $z \leqslant b'$, $SC_n(A, B)$ would not be positive, meaning that the events would not be correlated at stage $n$; were $z > b' + 1$, it would follow that $SC_n(A, B) > k$, which would mean the pair was already correlated at stage $n-1$.) Similarly, if $z = a'$, then $k = a' + 1$. Remember than by our convention we always have $a' \geqslant b'$. Finally, notice that if $a' \geqslant k$ and $b' \geqslant z$ then $SC_n(A, B)$ is negative and so there is no correlation; similarly if $b' \geqslant k$ and $a' \geqslant z$.

**Lemma 16.** *Suppose $A, B$ form a pair of logically independent events correlated at stage $n$. Suppose further that they have a common cause or a 0-type SCCS of size 3 at that stage. Then they have a common cause or a 0-type SCCS of size 3 at stage $n + 1$.*

*Proof.* (Note that the cases are not exclusive; they are, however, exhaustive, which is enough for the present purpose.)

**Case 1: $A, B$ have a 0-type common cause at stage $n$.**

Let $C$ be a 0-type common cause for the correlation. When moving from stage $n$ to $n+1$, a new atom $(n+1)$ is added. Let $C'^{\perp} = C^{\perp} \cup \{n+1\}$. Notice that $C$ and $C'^{\perp}$ form a partition of unity of the algebra at stage $n+1$. $C$ contains exclusively atoms from the algebra at stage $n$ and so continues to be a screener-off. Notice that since $C$ was a 0-type common cause at stage $n$, at that stage $P(A \mid C^{\perp}) = 0$ or $P(B \mid C^{\perp}) = 0$. Since the atom $n+1$ lies outside the events $A$ and $B$, at stage $n+1$ we have $P(A \mid C'^{\perp}) = 0$ or $P(B \mid C'^{\perp}) = 0$, and so $C'^{\perp}$ is a screener-off too. Thus $C$ and $C'^{\perp}$ are both screener-offs and compose a partition of unity at stage $n+1$. By corollary 4 (p. 34), this is enough to conclude that $A, B$ have a 0-type common cause at stage $n+1$.

**Case 2: $A, B$ have a common cause which is not a 0-type common cause at stage $n$.**

Let $C$ be a non-0-type common cause for the correlation at stage $n$. Notice that both $P(AB \mid C)$ and $P(AB \mid C^{\perp})$ are non-zero. In this case the "new" atom cannot be added to $C$ or $C^{\perp}$ without breaking the corresponding screening off condition. However—as we remarked in the previous case—the atom $n+1$ lies outside the events $A$ and $B$, so the singleton $\{n+1\}$ is trivially a screener-off for the pair. Since conditioning on $\{n+1\}$ gives probability 0 for both $A$ and $B$, the statistical relevance condition is satisfied. Therefore our explanation of the correlation at stage $n+1$ will be a 0-type SCCS of size 3: $C' = \{C, C^{\perp}, \{n+1\}\}$.[6]

**Case 3: $A, B$ have a 0-type SCCS of size 3 at stage $n$.**

Let the partition $C = \{C_i\}_{i \in \{0,1,2\}}$ be a 0-type SCCS of size 3 at stage $n$ for the correlation, with $C_2$ being the zero element (that is $P(A \mid C_2) = 0$ or $P(B \mid C_2) = 0$ (or possibly both), with the conditional probabilities involving $C_0$ and $C_1$ being positive). Let $C' = \{C_0, C_1, C_2 \cup \{n+1\}\}$. Appending the additional atom to $C_2$ does not change any conditional probabilities involved, so the statistical relevance condition is satisfied. Since $n+1 \notin A \cup B$, $C_2 \cup \{n+1\}$ screens off the correlation at stage $n+1$ and $C'$ is a 0-type SCCS of size 3 at stage $n+1$ for the correlation. □

As mentioned above, lemmas 14–16 complete the proof of this direction of the theorem since a method is given for obtaining a statistical common cause or an SCCS of size 3 for any correlation between logically independent events in any finite probability space with uniform distribution.

We proceed with the proof of the "upward" direction of theorem 4.

**Causal up-to-3-closedness w.r.t. $L_{ind}$ ⇒ Measure uniformity**

In fact, we will prove the contrapositive: if in a finite probability space with no 0-probability atoms the measure is not uniform then there exist logically independent,

6 The fact that a correlation has an SCCS of size 3 does not necessarily mean it has no common causes.

correlated events $A, B$ possessing neither a common cause nor an SCCS of size 3.[7] In the remainder of the proof we extend the reasoning from case 2 of proposition 4 of Gyenis and Rédei (2004), which covers the case of common causes.

Consider the space with $n$ atoms; arrange the atoms in the order of decreasing probability and label them as numbers $0, 1, \ldots, n-1$. Let $A = \{0, n-1\}$ and $B = \{0, n-2\}$. Gyenis and Rédei (2004) prove that $A, B$ are correlated and do not have a common cause. We will now show that they do not have an SCCS of size 3 either.

Suppose $C = \{C_i\}_{i\in\{0,1,2\}}$ is an SCCS of size 3 for the pair $A, B$. If for some $i \in \{0, 1, 2\}$ $A \subseteq C_i$ then $C$ violates the statistical relevance condition, since for the remaining $j, k \in \{0, 1, 2\}, j \neq k, i \neq j, i \neq k$, $P(A \mid C_j) = 0 = P(A \mid C_k)$. Similarly if $B$ is substituted for $A$ in the above reasoning. It follows that none of the elements of $C$ can contain the whole event $A$ or $B$. Notice also that no $C_i$ can contain the atoms $n-1$ and $n-2$, but not the atom 0, as then it would not be a screener-off. This is because in such a case $P(AB \mid C_i) = 0$ despite the fact that $P(A \mid C_i) \neq 0$ and $P(B \mid C_i) \neq 0$. But since $C$ is a partition of unity of the space, each of the three atoms forming $A \cup B$ has to belong to an element of $C$, and so each $C_i$ contains exactly one atom from $A \cup B$. Therefore for some $j, k \in \{0, 1, 2\}$ $P(A \mid C_j) > P(A \mid C_k)$ but $P(B \mid C_j) < P(B \mid C_k)$, which means that $C$ violates the statistical relevance condition. All options exhausted, we conclude that the pair $A, B$ does not have an SCCS of size 3; thus the probability space is not causally up-to-3-closed. □

The reasoning from the "upward" direction of the theorem can be extended to show that if a probability space with no 0-probability atoms has a non-uniform probability measure it is not causally up-to-$n$-closed for any $n \geqslant 2$. The union of the two events $A$ and $B$ described above only contains 3 atoms; it follows that the pair cannot have an SCCS of size greater than 3 since it would have to violate the statistical relevance condition (two or more of its elements would, when conditioned upon, give probability 0 to event $A$ or $B$). This, together with proposition 3 of Gyenis and Rédei (2004), justifies the following claims:

**Theorem 5.** *No finite probability space with a non-uniform measure and without 0-probability atoms is causally up-to-n-closed w.r.t. $L_{ind}$ for any $n \geqslant 2$.*

**Corollary 17.** *No finite probability space with a non-uniform measure and without 0-probability atoms is causally n-closed w.r.t. $L_{ind}$ for any $n \geqslant 2$.*

The proofs of lemmas 11 and 12 in section 5.3.3 will make it clear how to generalize both theorem 5 and corollary 17 to arbitrary finite spaces (also those possessing some 0-probability atoms) with a non-uniform measure. We omit the tedious details.

7 Recall that by assumption the probability space under consideration has at least 4 atoms.

#### 5.3.2.1 Examples

We will now present a few examples of how our method of finding explanations for correlations works, analyzing a few cases of correlated logically independent events in probability spaces of various sizes (with uniform probability distribution).

**Example 3** ($n = 7, A = \{0, 2, 3, 5, 6\}, B = \{1, 2, 5, 6\}$).

We see that $a' = 2$, $b' = 1$ and $k = 3$, so we will analyze the pair $A_1 = \{0, 1, 2, 3, 4\}$, $B_1 = \{2, 3, 4, 5\}$. We now check whether $A_1$ and $B_1$ were independent at stage 6, and since at that stage $A_1^\perp \cap B_1^\perp = \emptyset$ we conclude that they were not. Therefore the pair $A_1, B_1$ appears from above at stage 7. Notice that $SC_7(A_1, B_1) = 1$. By construction from lemma 14 we know that an event consisting of just a single atom from the intersection of the two events satisfies the requirements for being a common cause of the correlation. Therefore $C = \{2\}$ is a common cause of the correlation between $A$ and $B$ at stage 7.

**Example 4** ($n = 10, A = \{2, 3, 8\}, B = \{2, 8, 9\}$).

We see that $a' = 1$, $b' = 1$ and $k = 2$, so we will analyze the pair $A_1 = \{0, 1, 2\}$, $B_1 = \{1, 2, 3\}$. Since $SC_{10}(A_1, B_1) = 11$, we conclude that the lowest stage at which the pair is correlated is 5 (as remarked earlier, $SC$ changes by $k$ from stage to stage). $A_1$ and $B_1$ are logically independent at that stage, but not at stage 4, which means that the pair appears from above at stage 5. We employ the same method as in the previous example to come up with a 1-type common cause of the correlation at that stage—let it be the event $\{1\}$. Now the reasoning from case 2 of lemma 16 is used to "translate" the explanation to stage 6, where it becomes the following 0-type SCCS: $\{\{1\}, \{0, 2, 3, 4\}, \{5\}\}$. Case 3 of the same lemma allows us to arrive at an SCCS for $A_1$,$B_1$ at stage 10: $\{\{1\}, \{0, 2, 3, 4\}, \{5, 6, 7, 8, 9\}\}$. Its structure is as follows: one element contains a single atom from the intersection of the two events, another the remainder of $A_1 \cup B_1$ as well as one atom not belonging to any of the two events, while the third element of the SCCS contains the rest of the atoms of the algebra at stage 10. We can therefore produce a 0-type SCCS for $A$ and $B$ at stage 10: $\{\{2\}, \{0, 3, 8, 9\}, \{1, 4, 5, 6, 7\}\}$.

**Example 5** ($n = 12, A = \{2, 4, 6, 8, 9, 10, 11\}, B = \{1, 3, 6, 10, 11\}$).

We see that $a' = 4$, $b' = 2$ and $k = 3$, so we will analyze the pair $A_1 = \{0, 1, 2, 3, 4, 5, 6\}$, $B_1 = \{4, 5, 6, 7, 8\}$. We also see that $A_1$ and $B_1$ were logically independent at stage 11, but were not correlated at that stage. Therefore the pair $A_1, B_1$ appears from below at stage 12. Notice that $z = 3$. Therefore we see that $z > b'$ and $a' > k$, which means we can use the method from case 2 of lemma 15 to construct a 0-type common cause, whose complement consists of 1 atom from $A_1 \setminus B_1$ and 1 atom from $(A_1 \cup B_1)^\perp$. Going back to $A$ and $B$, we see that the role of the complement of our common cause can be fulfilled by $C^\perp = \{0, 2\}$. Therefore

$C = \{1, 3, 4, 5, 6, 7, 8, 9, 10, 11\}$ is a 0-type common cause of the correlation between $A$ and $B$ at stage 12.[8]

### 5.3.3 Proof of Lemmas 10-12

*Proof.* **[of lemma 10]** If $P$ is uniform, then $\langle S, P\rangle$ has no 0-probability atoms, which means that $S = S^+$ and $P = P^+$. Therefore $P^+$ is uniform, so (by theorem 3) $\langle S^+, P^+\rangle$ (and, consequently, $\langle S, P\rangle$) is causally up-to-3-closed w.r.t. $L^+_{ind}$. But in a space with no 0-probability atoms $L_{ind} = L^+_{ind}$, therefore $\langle S, P\rangle$ is also causally up-to-3-closed w.r.t. $L_{ind}$. □

The next two proofs will require "jumping" from $\langle S^+, P^+\rangle$ to $\langle S, P\rangle$ and *vice versa*. We will now have to be careful about the distiction between proper and improper SCC(S)s. Some preliminary remarks are in order.

Let $A \in S$. As before, we can think of $A$ as a set of atoms of $S$. Let $A^+$ be the set of non-zero probability atoms in $A$:

$$A^+ := A \setminus \{a \mid a \text{ is an atom of } S \text{ and } P(a) = 0\}.$$

Notice that

$$P(A) = \sum_{a \in A} P(a) = \sum_{a \in A^+} P(a) = P(A^+) = P^+(A^+). \tag{5.1}$$

Suppose $A, B, C \in S$. From (5.1) it follows that if $A, B$ are correlated in $\langle S, P\rangle$, $A^+, B^+$ are correlated in $\langle S^+, P^+\rangle$. Similarly, for any $D \in S$, $P(D \mid C) = P^+(D^+ \mid C^+)$. So, if $C$ screens off the correlated events $A, B$ in $\langle S, P\rangle$, then $C^+$ screens off the correlated events $A^+, B^+$ in $\langle S^+, P^+\rangle$. Also, if a family $\mathbf{C} = \{C_i\}_{i \in I}$ satisfies the statistical relevance condition w.r.t. $A, B$ in $\langle S, P\rangle$ then the family $\mathbf{C}^+ = \{C_i^+\}_{i \in I}$ satisfies the statistical relevance condition w.r.t. $A^+, B^+$ in $\langle S^+, P^+\rangle$. If $\mathbf{C} = \{C_i\}_{i \in \{0,\ldots,n-1\}}$ is a proper SCCS of size $n$ for the correlation between events $A, B$ in $\langle S, P\rangle$ then all its elements differ from both $A$ and $B$ by more than a measure zero event. It follows that in such a case $\mathbf{C}^+ = \{C_i^+\}_{i \in \{0,\ldots,n-1\}}$ is a proper SCCS of size $n$ for the correlation between events $A^+, B^+$ in $\langle S^+, P^+\rangle$.

*Proof.* **[of lemma 11]** Since $P^+$ is not uniform, by theorem 3 $\langle S^+, P^+\rangle$ is not causally up-to-3-closed w.r.t. $L^+_{ind}$ (and, consequently, $L_{ind}$). Then there exist logically independent, correlated events $A^+, B^+$ in $S^+$ which do not have a proper SCCS of size at most 3 in $\langle S^+, P^+\rangle$. The two events are also logically independent and correlated in $\langle S, P\rangle$; it

8 Incidentally, if we wanted to find a 1-type common cause for $A$ and $B$ at stage 12, we could put $C = \{2, 11\}$, in which case $P(A \mid C) = 1$. However, this is not always possible and there are cases in which only 0-type common causes (or only 1-type common causes) are possible. For a concrete example, take the pair $\{\{0, 1, 2, 3, 4\}, \{4, 5\}\}$, which appears from below at stage 11 and has only 0-type common causes at that stage (we wrote a computer program to verify this).

is easy to show that in $\langle S, P\rangle$ the pair $\langle A^+, B^+\rangle$ also belongs both to $L^+_{ind}$ and to $L_{ind}$. We will show that $\langle S, P\rangle$ also contains no proper SCCS of size at most 3 for these events. For suppose that for some $m \in \{2,3\}$, $\mathbf{C} = \{C_i\}_{i\in\mathbb{N},i<m}$ was a proper SCCS of size $m$ for the correlation between $A^+$ and $B^+$ in $\langle S, P\rangle$. Then $\mathbf{C}^+ := \{C_i^+\}_{i\in\mathbb{N},i<m}$ would be a proper SCCS of size $m$ for the correlation between $A^+$ and $B^+$ in $\langle S^+, P^+\rangle$, but by our assumption no such SCCSs exist. We infer that the correlated events $A^+, B^+$ have no proper SCCS of size up to 3 in $\langle S, P\rangle$, so the space $\langle S, P\rangle$ is not causally up-to-3-closed w.r.t. either $L_{ind}$ or $L^+_{ind}$. □

*Proof.* **[of lemma 12]** Since $P^+$ is uniform, by theorem 3 $\langle S^+, P^+\rangle$ is causally up-to-3-closed w.r.t. $L^+_{ind}$. We will first show that also $\langle S, P\rangle$ is causally up-to-3-closed w.r.t. $L^+_{ind}$. Notice that if $A, B \in S$ are correlated in $\langle S, P\rangle$ and $\langle A, B\rangle \in L^+_{ind}$ then $A^+, B^+ \in S^+$ are correlated in $\langle S^+, P^+\rangle$ and $\langle A^+, B^+\rangle \in L^+_{ind}$. We know that in that case there exists in $\langle S^+, P^+\rangle$ a proper SCCS of size 2 or 3 for $A^+$ and $B^+$. If we add the 0-probability atoms of $S$ to one of the elements of the SCCS, we arrive at a proper SCCS of size 2 or 3 for $A, B \in S$.

It remains to consider correlated events $A, B \in S$ such that $\langle A, B\rangle \in L_{ind}$ but $\langle A, B\rangle \notin L^+_{ind}$. In such a case at least one of the probabilities from definition 18 has to be equal to 0. It is easy to show that, since we know the two events are correlated, it can only be the case that $P(A \cap B^\perp) = 0$ or $P(B \cap A^\perp) = 0$; equivalently, $A^+ \subseteq B^+$ or $B^+ \subseteq A^+$. It may happen that $A^+ = B^+$. Let us first deal with the case of a strict inclusion; suppose without loss of generality that $A^+ \subset B^+$. If $|B^+ \setminus A^+| > 1$, take an event $C$ such that $A^+ \subset C \subset B^+$. Since both inclusions in the last formula are strict, in such a case $C$ is a proper statistical common cause for $A$ and $B$. Notice that since $\langle A, B\rangle \in L_{ind}$, from the fact that $A^+ \subset B^+$ it follows that $A \neq A^+$. Therefore, if $|B^+ \setminus A^+| = 1$, put $C = A^+$. Such a $C$ is an improper statistical common cause of $A$ and $B$.

The last case is that in which $A^+ = B^+$. From the fact that $A$ and $B$ are logically independent it follows that $A \setminus B^+ \neq \emptyset$ and $B \setminus A^+ \neq \emptyset$. Therefore $A \neq A^+$ and $B \neq B^+$. We can thus put $C = A^+$ or $C = B^+$ to arrive at an improper statistical common cause of $A$ and $B$.

When $A^+ \subseteq B^+$, it is also impossible to find (even improper) SCCSs of size 3 for $A$ and $B$. For suppose $\mathbf{C} = \{C_i\}_{i\in\{0,1,2\}}$ was an SCCS for $A$ and $B$. If for some $j \neq l; j, l \in \{0,1,2\}$ it is true that $C_j \cap A^+ = C_l \cap A^+ = \emptyset$, then $P(A|C_j) = 0 = P(A|C_l)$ and so $\mathbf{C}$ cannot be an SCCS of $A$ and $B$ due to the statistical relevance condition being violated. Thus at least two elements of $\mathbf{C}$ have to have a nonempty intersection with $A^+$. Every such element $C_j$ screens off $A$ from $B$. Since by our assumption $A^+ \subseteq B^+$, it follows that $P(AB|C_j) = P(A|C_j)$. Therefore the screening off condition takes the form of $P(A|C_j) = P(A|C_j)P(B|C_j)$; and so $P(B|C_j) = 1$. Since we already established that $C$ contains at least two elements which can play the role of $C_j$ in the last reasoning, it follows that in this case the statistical relevance condition is violated too; all options exhausted, we conclude that no SCCSs of size 3 exist for $A$ and $B$ when $A^+ \subseteq B^+$. The argument from this paragraph can also be applied to show that if $A^+ \subseteq B^+$ and $|B^+ \setminus A^+| \leqslant 1$, no proper statistical common causes for the two events exist. □

## 5.4 The "Proper" / "Improper" Common Cause Distinction and the Relations of Logical Independence

A motivating intuition for the distinction between proper and improper common causes is that a correlation between two events should be explained by a *different* event. The difference between an event $A$ and a cause $C$ can manifest itself on two levels: the algebraical ($A$ and $C$ being not identical as elements of the event space) and the probabilistic ($P(A \cap C^\perp)$ or $P(C \cap A^\perp)$ being not equal to 0). As per definition 19, in the case of improper common causes the difference between them and at least one of the correlated events (say, $A$) is only algebraical. For some this is intuitively enough to dismiss $C$ as an explanation for any correlation involving $A$.

One could, however, have intuitions to the contrary. First, events which differ by a measure zero event can be conceptually distinct. Second, atoms with probability 0 should perhaps be irrelevant when it comes to causal features of the particular probability space, especially when the independence relation considered is defined without any reference to probability. If the space is causally up-to-$n$-closed w.r.t. $L_{ind}$, adding 0-probability atoms should not change its status. But consider what happens when we add a single 0-probability atom to a space which is up-to-2-closed (common cause closed) w.r.t. $L_{ind}$. Take the space $\langle S_5, P_u \rangle$, where $S_5$ is the Boolean algebra with 5 atoms $\{0, 1, \dots, 4\}$ and $P_u$ is the uniform measure on $S_5$. This space is common cause closed by Proposition 3 from Gyenis and Rédei (2004). Label the added 0-probability atom as the number 5. It is easy to check that the pair $\langle \{3, 4\}, \{4, 5\} \rangle$ belongs to $L_{ind}$, is correlated and has no proper common cause. The only common cause for these events, $\{4\}$, is improper. Therefore the space is not common cause closed w.r.t. $L_{ind}$ in the sense of Gyenis and Rédei (2004) and our definition 20; this change in the space's status has been accomplished by adding a single atom with probability 0.

It should be observed that the pair of events belongs to $L_{ind}$, but not to $L^+_{ind}$; and that the bigger space is still common cause closed, but with respect to $L^+_{ind}$, not $L_{ind}$.

In general, suppose $\langle S, P \rangle$ is a space without any 0-probability atoms, causally up-to-$n$-closed w.r.t. $L_{ind}$, and suppose some "extra" atoms were added, so that a new space $\langle S', P' \rangle$ is obtained, where for any atom $a$ of $S'$,

$$P'(a) = \begin{cases} P(a) & \text{for } a \in S \\ 0 & \text{for } a \in S' - S. \end{cases}$$

It is easy to prove, using the techniques employed in the proof of lemma 12, that all "new" correlated pairs in $\langle S', P' \rangle$ belonging to $L_{ind}$ have (sometimes only improper) SCCSs of size up to $n$. This is also true in the special case of $\langle S_5, P_u \rangle$ augmented with some 0-probability atoms. Perhaps, then, we should omit the word "proper" from the requirements for a probability space to be causally up-to-$n$-closed (definition 20)?

This, however, is only one half of the story. Suppose the definition of causal up-to-$n$-closedness were relaxed in the above way, so that explaining correlations by means of improper SCC(S)s would be admissible. Consider a space $\langle S^+, P^+\rangle$,[9] in which $S^+$ has at least 4 atoms and $P^+$ is not the uniform measure on $S^+$. This space, as we know, is not causally up-to-3 closed in the sense of definition 20, but it is also not causally up-to-3 closed in the "relaxed" sense, since the difference between proper and improper common causes can only manifest itself in spaces with 0-probability atoms.[10] When a new 0-probability atom $m$ is added every hitherto unexplained correlation between some events $A$ and $B$ gains an SCC by means of the event $C := A \cup \{m\}$. All such SCCs are, of course, improper.

In short, the situation is this: if proper SCC(S)s are required this leads to somewhat unintuitive consequences regarding causal up-to-$n$-closedness w.r.t. $L_{ind}$. Omitting the requirement results, however, in unfortunate effects regarding causal up-to-$n$-closedness no matter whether $L_{ind}$ or $L^+_{ind}$ is considered. We think the natural solution is to keep the requirement of proper SCC(S)s in the definition of causal up-to-$n$-closedness but, of the two independence relations, regard $L^+_{ind}$ as more interesting. It is the rightmost column of table 5.1 that contains the most important results of this chapter, then; this is fortunate, since they are a "pure" implication and an equivalence, without any special disclaimers.

## 5.5 Other Independence Relations

So far, the relation of independence under consideration—determining which correlations between two events require explanation—was the relation of logical independence and its derivative $L^+_{ind}$. Let us consider using a "broader" relation $R_{ind} \supset L_{ind}$, which apart from all pairs of logically independent events would also include some pairs of logically dependent events. (The spaces under consideration are still finite.) For clarity, assume the space does not have any 0-probability atoms (so that e.g. $L_{ind} = L^+_{ind}$), but make no assumptions regarding the uniformity of the measure. Will we have more correlations to explain? If so, will they have common causes?

First, observe that if $A$ or $B$ equals $\mathbf{1}_S$, and so $P(A)$ or $P(B)$ equals 1, there is no correlation. In the sequel assume that neither $A$ nor $B$ equals $\mathbf{1}_S$.

Second, note that if $A \cap B = \emptyset$, then $P(AB) = 0$ and no (positive) correlation arises.

9 Remember that by our convention such a space has no 0-probability atoms.

10 This is because the spaces we are dealing with are finite—so that we can be sure the Boolean algebras considered have atoms at all—and we already require an SCC for two events $A$ and $B$ to be distinct from both $A$ and $B$, see definition 9, p. 9.

Third, if $A^\perp \cap B^\perp = \emptyset$, there is again no positive correlation. This is because in such a case $P(AB) + P(AB^\perp) + P(A^\perp B) = 1$, and since $P(A)P(B) = P(AB)[P(AB) + P(AB^\perp) + P(A^\perp B)] + P(AB^\perp)P(A^\perp B) \geqslant P(AB)$, the events are not correlated.

Consider the last possible configuration in which the events $A, B$ are logically dependent: namely, that one is a subset of the other. Suppose $A \subseteq B$. Since by our assumption both $P(A)$ and $P(B)$ are strictly less than 1, the events will be correlated. It can easily be checked[11] that when $A \subseteq B$ but $B \neq \mathbf{1}_S$ any $C$ which screens off the correlation and has a nonempty intersection with $A$ (and so $P(A \mid C) \neq 0$) has to be a subset of $B$ (because $P(B \mid C) = 1$). And since it cannot be that both $C$ and $C^\perp$ are subsets of $B$, then if $C$ is a common cause, it is necessary that $C^\perp \cap A = \emptyset$. In the other direction, it is evident that if $A \subseteq C \subseteq B$, both $C$ and $C^\perp$ screen off the correlation and the statistical relevance condition is satisfied. The only pitfall is that the definition of a common cause requires it be distinct from both $A$ and $B$, and so none exist when $b' = 1$.

To summarize, the only correlated pairs of logically dependent events $A, B$ are those in which one of the events is included in the other. Assume $A \subseteq B$. Then:

- if $b' = 1$, there is no common cause of the correlation;
- otherwise the common causes of the correlation are precisely all the events $C$ such that $A \subset C \subset B$.

Lastly, notice that in a space $\langle S_n, P_u\rangle$ ($S_n$ being the Boolean algebra with $n$ atoms and $P_u$ being the uniform measure) we could proceed in the opposite direction and restrict rather than broaden the relation $L_{ind}$. If we take the independence relation $R_{ind}$ to be the relation of logical independence restricted to the pairs which appear from above or below at stage $n$ then our probability space is common cause closed w.r.t. $R_{ind}$.

## 5.6 A Slight Generalization

In this section we will show that the results of this chapter, which have only concerned classical probability spaces so far, are also meaningful for finite non-classical spaces. We go back to our former practice: by "common cause" we will always mean "proper common cause"; similarly with "common cause system".

**Definition 24 [Non-classical probability space]** A lattice $L$ is *orthomodular* if $\forall_{e,f \in L}\ e \leqslant f \Rightarrow f = e \vee (e^\perp \wedge f)$.

Two elements $e$ and $f$ of $L$ are *orthogonal* iff $e \leqslant f^\perp$.

11 See the last paragraph of the proof of lemma 12, p. 85.

An *additive state* on an orthomodular lattice (OML) $L$ is a map $P$ from $L$ to $[0, 1]$ such that $P(\mathbf{1}_L) = 1$ and for any $E \subseteq L$ such that $E$ consists of mutually orthogonal elements, if $\bigvee E$ exists, then $P(\bigvee E) = \sum_{e \in E} P(e)$.[12]

A *non-classical probability space* is a pair $\langle L, P \rangle$, where $L$ is a non-distributive OML and $P$ is an additive state on $L$.[13]

A relation of compatibility needs to be introduced. Only compatible events may be correlated and a common cause needs to be compatible with both effects. We use the word "compatibility" because it was the one used in (Hofer-Szabó et al. (2000)); sometimes "commutativity" is used in its place (see e.g. Kalmbach (1983)).

**Definition 25 [Compatibility, correlation, SCC(S) in non-classical spaces]** Let $L$ be an OML and $e, f \in L$. Event $e$ is said to be *compatible* with $f$ ($eCf$) if $e = (e \wedge f) \vee (e \wedge f^{\perp})$.

Events $e, f$ are said to be *correlated* if $eCf$ and the events are correlated in the sense of definition 3.

The event $c \in L$ is a *proper statistical common cause* of $e$ and $f$ if it fulfills the four requirements from definition 9 (p. 30), differs from both $e$ and $f$ by more than a measure zero event, and is compatible both with $e$ and with $f$ (of course, $c^{\perp}$ will be compatible, too).

A partition $\{C_i\}_{i \in I}$ of $\mathbf{1}_L$ is a *proper statistical common cause system of size n* of $e$ and $f$ if it satisfies the requirements of definition 10 (p. 30), all its elements differ from both $e$ and $f$ by more than a measure zero event, and all its elements are compatible both with $e$ and $f$.

The notion of causal up-to-$n$-closedness is then immediately transferred to the context of non-classical probability spaces by substituting "non-classical" for "classical" in definition 20 (p. 73).

A *block* of an OML is its maximal Boolean subalgebra. We are interested in pairs of correlated events in $L$; since events are compatible iff they lie in a block (Kalmbach (1983), p. 39), it turns out that they can only be correlated if they belong to the same block.

---

12 Of course, in the finite case—since a lattice always contains all suprema of doubletons by virtue of being a lattice—it would suffice to say that for any two orthogonal elements $e$ and $f$, $P(e \vee f) = P(e) + P(f)$. However, infinite lattices can be *incomplete*: they can lack the suprema of certain subsets.

13 A different direction could be taken in presenting the definitions of classical and non-classical probability spaces: first, a probability space could be defined as a measure on an OML; then, classical and non-classical spaces could be distinguished on the basis of whether the OML in question is distributive (in which case it is by definition a Boolean algebra) or not. However, throughout the biggest part of this essay we have been in the sphere of classical probability and so the term "probability space" was effectively short for "classical probability space". We do not want to change this more than halfway through the work. However, for clarificatory reasons, this will result in some definitions containing the phrase "classical or non-classical", which is perhaps redundant at first sight—see e.g. definitions 26 and 27 below.

This leads us to the result of this section, which can be colloquially phrased in this way: a finite non-classical probability space is causally up-to-$n$ closed if and only if all its blocks are causally up-to-$n$-closed.

**Theorem 6.** *Suppose $\langle L, P\rangle$ is a finite non-classical probability space. Suppose all blocks of $L$ have at least 4 atoms $e$ such that $P(e) > 0$. Then $\langle L, P\rangle$ is causally up-to-$n$-closed w.r.t $L_{ind}$ if and only if for any block $B$ of $L$, the classical probability space $\langle B, P|_B\rangle$ is causally up-to-$n$-closed w.r.t. $L_{ind}$.*

*Proof.* Suppose $\langle L, P\rangle$ is causally up-to-$n$-closed w.r.t. $L_{ind}$. Let $B$ be a block of $L$; let $e, f$ be correlated and logically independent events in $\langle B, P|_B\rangle$. Then $e, f$ are correlated and logically independent events in $\langle L, P\rangle$, and so have an SCCS of size up to $n$ in $\langle L, P\rangle$. But since all elements of the SCCS have to be compatible with $e$ and $f$, they also have to belong to $B$. And so the pair has an SCCS of size up to-$n$ in $\langle B, P|_B\rangle$.

For the other direction, suppose that for any block $B$ of $L$ the space $\langle B, P|_B\rangle$ is causally up-to-$n$-closed w.r.t. $L_{ind}$. Let $e, f$ be correlated and logically independent events in $\langle L, P\rangle$. Being correlated entails being compatible; and so $e$ and $f$ belong to a block $B$. Since the ordering on $L$ is induced by the orderings of the elements of $B$, $e$ and $f$ are also logically independent in $B$. Therefore by our assumption they have an SCCS of size up to $n$ in $\langle B, P|_B\rangle$. This SCCS is a partition of unity of $L$, and so satisfies definition 25. Thus $e$ and $f$ have an SCCS of size up to $n$ in $\langle L, P\rangle$. □

### 5.6.1 Examples

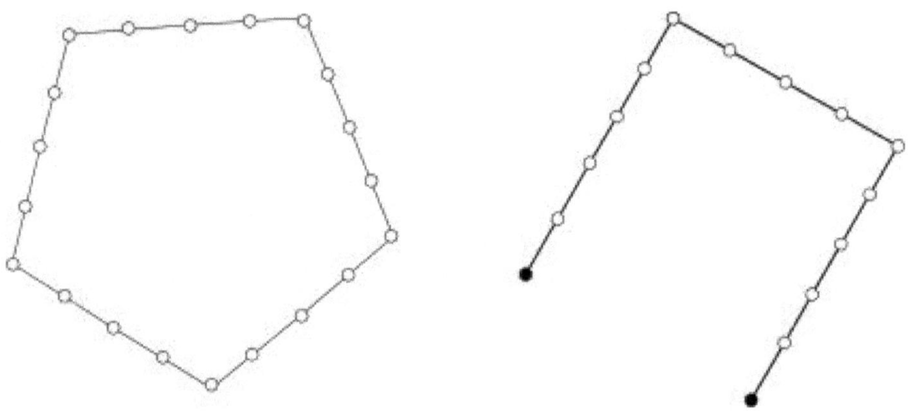

**Figure 5.1:** Greechie diagrams of two OMLs which, if supplied with the state which assigns number $\frac{1}{5}$ to all "white" atoms and 0 to both "black" atoms, form non-classical probability spaces which are causally up-to-2-closed (or simply "causally closed", to use the term of Gyenis and Rédei (2004)) w.r.t. $L^+_{ind}$.

We will now present a few examples of causal closedness and up-to-3-closedness of non-classical probability spaces. Figure 1 depicts two non-classical probability spaces causally closed w.r.t. $L^+_{ind}$. Notice that all blocks have exactly 5 atoms of non-zero probability and each such atom receives probability $\frac{1}{5}$, and so each block is causally closed w.r.t. $L^+_{ind}$. The left space is also causally closed w.r.t. $L_{ind}$.

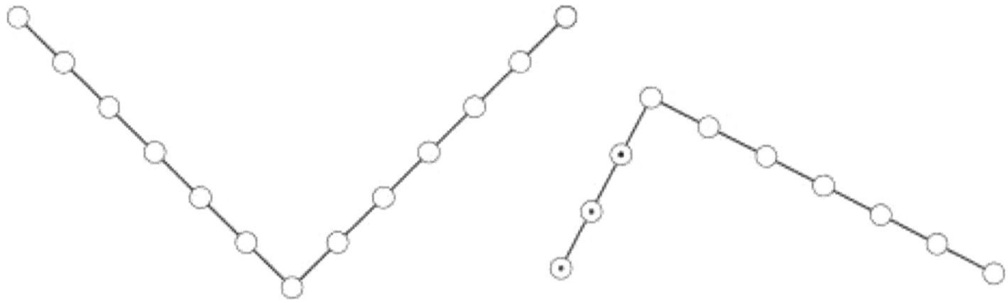

**Figure 5.2:** In these OMLs "white" atoms have probability $\frac{1}{7}$ and the "dotted" ones $\frac{2}{7}$. The space depicted on the left is causally up-to-3-closed, but the one on the right is not.

The left OML in figure 2 has two blocks and the measure of the space is uniform on both of them. Therefore the space is causally up-to-3-closed w.r.t. $L_{ind}$. This, however, is not the case with the right one: its measure is not uniform on the block with four atoms. There is then a correlation among some two logically independent events from that block which has neither a common cause nor an SCCS of size 3. (One of these events will contain one "dotted" atom and the single "white" atom of the block; the other will contain two "dotted" atoms.) Therefore the space is not causally up-to-3-closed w.r.t. $L_{ind}$.

## 5.7 Application for Constructing Bayesian Networks—a Negative Opinion

One could entertain the thought that the algorithm outlined above, which describes the construction of SCCs and SCCSs for pairs of logically independent, correlated events in finite classical probability spaces with the uniform measure, could be useful in the process of constructing a Bayesian network; the prospect seems to be encouraging since the networks consist of a finite number of variables with each having a finite set of possible values (Williamson (2005)). Suppose we have a fixed population on which we are doing our research but we are not sure what variables should belong to the causal structure, apart from some we are certain of. Suppose two correlated variables, $X$ and $Y$, belong to the group we are certain to have a rightful place in the structure but no other variable we are similarly sure of can be a common ancestor of

$X$ and $Y$ in our projected DAG, since no set of variables under consideration (different from $X$ and $Y$) makes $X$ and $Y$ independent when conditioned upon. Could the algorithm presented above allow us to construct an additional node for our Bayesian network, representing a variable we were not aware of when initially considering the causal structure in question which, when conditioned on, would render the correlated variables independent? Unfortunately, the answer is "no"; this will bring us back to the topic of common and common common screener systems.

The reason for the negative conclusion is the fact that a correlation between variables typically entails numerous correlations between events. The argument above provides an explanation for any single correlation between logically independent events, but there is no guarantee that a way exists of somehow "combining" the events (= common causes), or three element partitions of the population (= SCCSs), in order to create a random variable, which is what we need if we want to put a new node in our DAG.

A few simple abstract examples will clarify this. First, a "fortunate" case. Consider an 8-element universe $\Omega = \{0, 1, \ldots, 7\}$ and two correlated random variables $A$ and $B$, defined in this way:

$$A(x) = \begin{cases} 1, & x \in \{0, 1\}; \\ 2, & x \in \{2, 3\}; \\ 3, & x \in \{4, 5, 6, 7\}; \end{cases} \qquad B(x) = \begin{cases} 1, & x \in \{0, 3\}; \\ 2, & x \in \{1, 2\}; \\ 3, & x \in \{4, 5, 6, 7\}. \end{cases}$$

There are some cases of correlated events (assume the uniform measure on $\Omega$), e.g. $P(A = 1 \wedge B = 2) > P(A = 1)P(B = 2)$ and $P(A = 2 \wedge B = 2) > P(A = 2)P(B = 2)$. If we run the procedure outlined in the preceding sections we arrive at the event $C = \{0, 1, 2, 3\}$, which happens to be a common statistical common cause for all correlations. Therefore we can define it as a random variable:

$$C(x) = \begin{cases} 1, & x \in \{0, 1, 2, 3\}; \\ 2, & x \in \{4, 5, 6, 7\} \end{cases}$$

which may be a candidate for a node in our DAG, since it satisfies the requirement for a common ancestor of $A$ and $B$, making them independent when conditioned upon.[14] In fact, being an SCC is overkill here; it is evident that being a common common screener system suffices. Of course, as noted in chapter 2, trivial systems of this kind always exist but one could be hopeful that our procedure may generate non-trivial systems for subsequent consideration.

Unfortunately, in general this is not the case. Consider a smaller, 7-element universe $\Omega = \{0, 1, \ldots, 6\}$ and two correlated random variables $A$ and $B$, defined in this

---

14 More precisely, the requirement is that the set of *all* common ancestors should make the correlated variables independent when conditioned upon. By assumption, we do not have other common ancestors of the events in question in our graph.

way:

$$A(x) = \begin{cases} 1, & x \in \{0,1\}; \\ 2, & x \in \{2,3,4\}; \\ 3, & x \in \{5,6\}; \end{cases} \qquad B(x) = \begin{cases} 1, & x \in \{0,3,4\}; \\ 2, & x \in \{1,2\}; \\ 3, & x \in \{5,6\}. \end{cases}$$

The SCC $C$ suggested by our procedure for the correlation between events $A = 1$ and $B = 2$ consists of $0, 1, 2$ and one element from $\{3, 4\}$. Unfortunately, the SCC $D$ suggested by our procedure for the correlation between events $A = 2$ and $B = 1$ consists just of a single element from $\{3, 4\}$. Neither of the two SCCs is a common SCC for the two pairs of correlated events; a 4 element partition of $\Omega$ consisting of all Boolean combinations of $C$ and $D$ also fails to be a common SCCS, and even a common common screener system for the two correlated pairs. In cases like that, which we conjecture are more frequent than the "fortunate" ones from the previous paragraph, our procedure unfortunately does not yield the information needed to define a random variable which could be a candidate for a node in the DAG to be constructed.

## 5.8 The Existence of *Deductive Explanantes*

We will now prove a theorem concerning the existence of *deductive explanantes*, a notion introduced in section 2.4.2, for all correlations between logically independent events in finite classical probability spaces with the uniform measure.

**Theorem 7.** *Let $\langle S, P\rangle$ be a finite classical probability space with the uniform measure. Suppose $A$ and $B$ are correlated, logically independent events. Then there exists an event $C \in S$ which is a deductive explanans for $A$ and $B$.*

*Proof.* Again, without loss of generality we can consider $S$ as an $n$-atom Boolean algebra of all subsets of the set $\{0, \ldots, n-1\}$; in other words, as the probability space we would refer to as "stage $n$" in the preceding sections. The *deductive explanans* for $A$ and $B$ will be the common cause $C$ for them at the (possibly lower) stage at which they appear, either from above or from below.

Suppose that at stage $m \leqslant n$, where $A$ and $B$ appear, the method described in the proofs of lemmas 14 and 15 ascribes them a statistical common cause $C$. This $C$ of course also screens off $A$ in $B$ at stage $n$. However, at stage $n$ $C^{\perp}$ may be bigger then $C^{\perp}$ at stage $m$: the difference is $n - m$ atoms. $C^{\perp}$ screens off $A$ from $B$ at stage $m$. The addition of even a single atom to $C^{\perp}$ has to break the condition, since the atom has non-zero probability (due to the measure being uniform) and belongs to neither $A$ nor $B$. It will make $A$ and $B$ correlated conditional on $C^{\perp}$; at stage $m + 1$, $P(AB|C^{\perp}) > P(A|C^{\perp})P(B|C^{\perp})$.

To see this, recall lemma 13: if two events are independent at some stage, they are correlated at every later stage. Now consider a probability space where the event space is the Boolean algebra with the set of atoms consisting of the set of atoms of $B$ restricted to $C^{\perp}$ at stage $m$, and the measure is the corresponding restriction of $P$. This algebra is isomorphic to the space at stage $k$, where $k$ is the cardinality of $C^{\perp}$. Events $A$ and $B$ are independent at stage $k$ but due to lemma 13 are correlated at stage $k+1$. Thus, when an additional atom is appended to $C^{\perp}$, events $A$ and $B$ become correlated conditional on $C^{\perp}$. By the above argument, adding more atoms to $C^{\perp}$ does not change the fact that $A$ and $B$ are correlated conditional on it. And so, at stage $n$ it has to be the case that $P(AB|C^{\perp}) > P(A|C^{\perp})P(B|C^{\perp})$.

Since $C$ is a common cause of $A$ and $B$ at stage $m$, at that stage $P(A|C) > P(A|C^{\perp})$ and $P(B|C) > P(B|C^{\perp})$. When we move to higher stages, it is evident $P(A|C)$ and $P(B|C)$ stay the same, while both $P(A|C^{\perp})$ and $P(B|C^{\perp})$ decrease. Therefore STAT($A$,$C$) and STAT($B$,$C$) are true at stage $n$, too.

Lastly, it is just a matter of consulting inequality (2.3) on page 41 to see that $C$ is a *deductive explanans* for $A$ and $B$: the left-hand side of the inequality is positive, while the right-hand side is negative. □

So far, nothing has been established regarding the existence—or nonexistence—of *deductive explanantes* for events in finite spaces with non-uniform measures, or in infinite spaces.

## 5.9 Conclusions and Problems

The main result of this chapter is that in finite classical probability spaces with the uniform probability measure (and so no atoms with probability 0) all correlations between logically independent events have an explanation by means of a common cause or a common cause system of size 3. A few remarks are in order.

First, notice that the only SCCSs employed in our method described in section 5.3.2 are 0-type SCCSs, and that they are required only when “translating” the explanation from a smaller space to a bigger one. Sometimes (if the common cause we found in the smaller space is 0-type; see example 5 above) such a translation can succeed without invoking the notion of SCCS at all.

Second, #-type common causes, which some would view as “genuinely indeterministic”, are never *required* to explain a correlation – that is, a correlation can always be explained by means of a 0-type SCCS, a 0-type statistical common cause, or a 1-type statistical common cause[15]. Therefore one direction of the equivalence in theorem 4 can be strengthened:

15 But #-type common causes do exist: e.g. in the space with 12 atoms and the uniform measure the pair of events $\{A, B\}$, where $A = \{0,1,2,3,4,5,6\}$, $B = \{4,5,6,7,8\}$ (the same we dealt with in example 5, p. 83) has, apart from both 0- and 1-type common causes, a #-type common cause of shape $C = \{1,2,4,5,7,9\}$, $C^{\perp} = \{0,3,6,8,10,11\}$; $P(A \mid C) = \frac{2}{3}$, $P(B \mid C) = \frac{1}{2}$, $P(A \mid C^{\perp}) = \frac{1}{2}$, $P(B \mid C^{\perp}) = \frac{1}{3}$.

**Theorem 8.** *Let $\langle S, P\rangle$ be a finite classical probability space. Let $S^+$ be the subalgebra of $S$ containing all and only the non-zero probability atoms of $S$ and $P^+$ be the restriction of $P$ to $S^+$. Suppose $S^+$ has at least 4 atoms.*

*If $P^+$ is the uniform probability measure on $S^+$, then any pair of correlated and logically independent events in $\langle S, P\rangle$ has a 1-type statistical common cause, a 0-type statistical common cause or a 0-type statistical common cause system of size 3 in $\langle S, P\rangle$.*

## 5.10 Causal Closedness of Atomless Spaces

Let us recall an important theorem from Gyenis and Rédei (2004) which we will use in the next chapter. We move to the general context of possibly infinite probability spaces.

**Definition 26 [Atomless probability space]** A (classical or non-classical) probability space $\langle \mathcal{F}, \mu\rangle$ is *atomless* if for any $C \in \mathcal{F}$, if $\mu(C) > 0$, then there exists $D \in \mathcal{F}$ such that $D \subseteq C$ and $0 < \mu(D) < \mu(C)$.

Of course, an atomless space may consist of a measure defined on an atomic algebra (take the example of all Borel subsets of $[0, 1] \subseteq \mathbb{R}$ with the Lebesgue measure).

We will first focus on the classical case. It is obvious from the above definition that in any classical atomless space, for any non-zero measure event $C$ there is an infinite sequence of events with positive measure which are its subsets. It could be contemplated, though, that some real numbers lower than $\mu(C)$ are not exhibited as probabilities of events being subsets of $C$. This is impossible! Gyenis and Rédei (2004) use a fact concerning classical atomless spaces (see e.g. p. 46 of Fremlin (2001)) according to which for any $C$ such that $\mu(C) > 0$ and for any real number $r$ such that $0 < r < \mu(C)$ there exists a $D \in \mathcal{F}$ such that $D \subseteq C$ and $\mu(D) = r$. This allows the authors to prove the following fact:

**Fact 18** (Gyenis and Rédei (2004)). *All atomless classical probability spaces are causally closed w.r.t. $L_{ind}$.*

Kitajima (2008) extended Gyenis and Rédei's result to a special class of non-classical spaces: these in which the OML, on which the measure is defined, is atomless and complete (has suprema of all its subsets). Kitajima proves that in such a case the non-classical probability space is atomless, too, and that all such spaces contain SCCs for each pair of logically independent, correlated events. We state Kitajima's result in the following form:

**Fact 19** (Kitajima (2008)). *If in a non-classical probability space $\langle L, P\rangle$ $L$ is an atomless and complete OML then $\langle L, P\rangle$ is causally closed w.r.t. the relation of logical independence.*

# 6 Causal Completability

This chapter concerns the formulation of the PCC we have dubbed "PCC 4" (p. 9). The published results concerning causal completability (see e.g. Hofer-Szabó et al. (1999)) are always formulated in the "pair"-style of thinking about probability spaces: the one used in the previous chapter, according to which a probability space is an algebra-measure pair with no mention of a sample space. However, the main theorem of this chapter, theorem 10, due to its very nature has to be phrased in a way which uses sample spaces. We will therefore be switching from one way of writing to the other. Throughout the chapter we will write in the "pair"-style notation with the exception of section 6.2, when we will use the traditional, "triadic" notation.

The notion of an extension of a probability space was defined for probability spaces thought of as 3-tuples (definition 8, p. 9). We now repeat it in the "pair"-style, omitting the sample space, which allows us to switch to the more general formulation which is also applicable to non-classical spaces:

**Definition 27 [Extension]** A (classical or non-classical) probability space $\langle S', \mu' \rangle$ is called an *extension* of the probability space $\langle S, \mu \rangle$ iff there exists an orthomodular lattice embedding $h$ of $S$ into $S'$ such that for any $E \in S$, $\mu(E) = \mu'(h(E))$.

The idea of PCC 4 is that, even though there are some unexplained correlations in a given space, an extension of it might exist which would contain the required explanations. Such a case may represent a situation in which, initially, not all important factors are taken into account—and a more "fine-grained" probability space contains elements which do the job of explaining the correlations.

In Hofer-Szabó et al. (1999) the authors discuss causal completability with regard to a family of correlated pairs of events. Their main result regarding classical probability spaces follows, in a bit different formulation.

**Definition 28 [Causal completability]** Suppose $\langle S, P \rangle$ is a probability space and $\mathfrak{F}$ is a family of pairs of correlated events which do not have an SCC in $\langle S, P \rangle$. The space $\langle S, P \rangle$ is *causally completable with regard to the family* $\mathfrak{F}$ if there exists an extension $\langle S', P' \rangle$ of $\langle S, P \rangle$ by means of a homomorphism $h$ which contains an SCC for $\langle h(A), h(B) \rangle$ for every pair $\langle A, B \rangle \in \mathfrak{F}$.

**Fact 20** (Proposition 2 from Hofer-Szabó et al. (1999)). *Every classical probability space is causally completable with regard to any finite family of correlated events.*

Ref. Hofer-Szabó et al. (1999) pose the problem whether classical probability spaces are causally completable with regard to *infinite* families of correlated events. This we answer in the positive in section 6.2.

Of course, an extension of a given space which provides explanations for some correlations may very well introduce new unexplained correlations. The extension constructed by Hofer-Szabó et al. (1999) is not expected to be causally closed, or causally up-to-$n$-closed, for any natural number $n$. For a single unexplained correlation, the extension is made from two copies of the initial space (for details, see p. 391-392 of Hofer-Szabó et al. (1999)). In the next section we will present a simple method of extending probability spaces to spaces which are causally up-to-3-closed. The method will, however, be restricted to finite spaces with rational probabilities on the atoms.

## 6.1 Causal Completability the Easy Way—“Splitting the Atom”

If in a finite probability space with some non-uniform measure the atoms have rational probabilities we can “split” them into pieces in order to arrive at a space with the uniform measure which will be an obvious, intuitive extension of the initial space. Consider the following illustration:

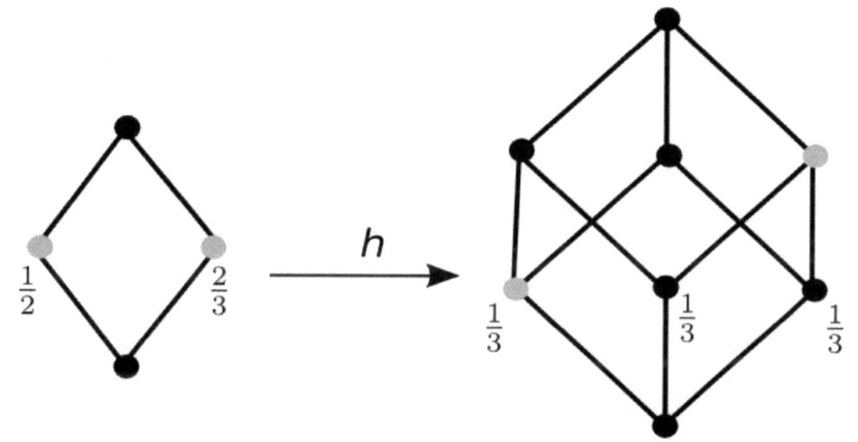


Here, we have a space with non-uniform measure on two atoms on the left, and its uniform-measure extension on the right. The image of one of the atoms of the left space through the homomorphism $h$ is still an atom, but the other atom (the one with probability $\frac{2}{3}$) loses this status; it is above two atoms in the extension.

Now, consider an example in which the initial space displays a correlated pair of events without an SCC (it is the same space which Hofer-Szabó et al. (2000) present as not causally closed):

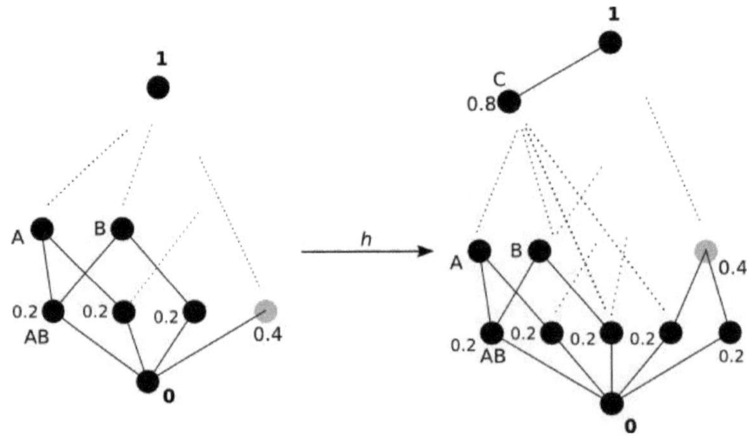


Events $A$ and $B$ are correlated in the left space but lack an SCC in that space. Their images in the space on the right, however, possess an SCC (the event $C$) the construction of which was possible due to the fact that the atom with probability $\frac{2}{5}$ has been split into two atoms with probability $\frac{1}{5}$. This is the most fortunate case possible, since the space on the right is (as we know from the results by Gyenis and Rédei (2004)) the only finite causally closed space. In general, the product of the procedure of "splitting the atoms" will be a finite space with uniform measure. But, explanation-wise, it is not a worse thing since (as we know from the results of the preceding chapter) all such spaces are causally up-to-3-closed (w.r.t. the relation of logical independence).

Let us put together the above considerations in form of a theorem.

**Theorem 9.** *Any finite probability space $\langle S, P\rangle$ with rational probabilities on the atoms of $S$ has an extension which is causally up-to-3-closed w.r.t. the relation of logical independence.*

**A sketch of the proof.** The probabilities of the atoms $\{a_0, \ldots, a_m\}$ of $S$ constitute a finite list $\langle p_0, \ldots, p_m\rangle$ of fractions. Calculate the lowest common denominator $D$ of these fractions. Let $S'$ be the Boolean algebra with $D$ atoms $\{b_1, \ldots, b_D\}$ and let $P'$ be the uniform measure on $S'$. Transform all the fractions $\langle p_0, \ldots, p_m\rangle$ so that their denominator is $D$. Let $\langle n_0, \ldots, n_m\rangle$ be the list of numerators of the corresponding fractions from $\langle p_0, \ldots, p_m\rangle$. Of course, $\sum_{i=0}^{m} n_i = D$. Let $h : S \to S'$ be a homomorphism which assigns any atom $a_i$ of $S$ the supremum of $n_i$ atoms of $S'$ in the following way:

$$\begin{aligned} h(a_0) &= \{b_1, \ldots, b_{n_0}\}; \\ h(a_1) &= \{b_{n_0+1}, b_{n_0+n_1}\}; \\ &\ldots \end{aligned}$$

It is evident that $\langle S', P' \rangle$ is an extension of $\langle S, P \rangle$ by means of the homomorphism $h$. And by theorem 4 (p. 75), this extension is causally up-to-3 closed w.r.t. the relation of logical independence.

## 6.2 Causal Completability of Classical Probability Spaces—the General Case

A way of solving the general problem of causal completability with regard to *any* family of correlated events would be to show that any space possesses a causally closed extension. A result of Gyenis and Rédei (2004), reproduced above as fact 18, states that all atomless spaces are causally closed. We simply need to find a way of extending an arbitrary space to an atomless space. This is done in the proof of theorem 10. Let us state the initial problem formally; it was posed in Hofer-Szabó et al. (1999) and Hofer-Szabó et al. (2000):

**Problem 4** (Causal completability of classical probability spaces). *Let $\langle S, \mu \rangle$ be a probability space and $\mathcal{W} \subseteq S^2$ be the (possibly infinite) family of all pairs of correlated logically independent events for which no common cause in $\langle S, \mu \rangle$ exists. Is there an extension $\langle S', \mu' \rangle$ (given by the embedding h) of $\langle S, \mu \rangle$ such that for any $\langle D, E \rangle \in \mathcal{W}$, there exists in $\langle S', \mu' \rangle$ a common cause for the pair $\langle h(D), h(E) \rangle$?*

In view of fact 18, we can answer this problem by showing that any classical probability space is extendable to an atomless space. We will use the following lemma:

**Lemma 21.** *Let $\langle S, \mathcal{F}, p \rangle$ be a probability space and let $\langle [0, 1], \mathcal{B}, L \rangle$ be the space of all Borel subsets of the* $[0, 1]$ *segment, L being the Lebesgue measure. Then the product space $\langle S \times [0, 1], \Sigma, \mu \rangle$ of the two above spaces is atomless.*

The proof uses the technique from chapter 211M of Fremlin (2001). For the details on the construction of $\Sigma$, the event $\sigma$-algebra of the product space, see e.g. chapter *IV*.6 of Feller (1968), vol. 2.

*Proof.* Let $E \in \Sigma$, $\mu(E) > 0$. Let $f$ be a function from $[0, \frac{1}{2}]$ to $[0, 1]$ given by the formula $f(a) = \mu\left(E \cap (\mathbf{1}_{\mathcal{F}} \times [\frac{1}{2} - a, \frac{1}{2} + a])\right)$. Observe that if $a, b \in [0, \frac{1}{2}]$ and $a \leqslant b$, then $f(a) \leqslant f(b) \leqslant f(a) + \mu(\mathbf{1}_{\mathcal{F}} \times [\frac{1}{2} - b, \frac{1}{2} + b]) - \mu(\mathbf{1}_{\mathcal{F}} \times [\frac{1}{2} - a, \frac{1}{2} + a]) = f(a) + p(\mathbf{1}_{\mathcal{F}}) \cdot L([\frac{1}{2} - b, \frac{1}{2} + b]) - p(\mathbf{1}_{\mathcal{F}}) \cdot L([\frac{1}{2} - a, \frac{1}{2} + a]) = f(a) + 2b - 2a$. Therefore, $f$ is continuous (as $b$ approaches $a$, $f(b)$ approaches $f(a)$).

Notice that $f(0) = 0$ and $\lim_{n \to \frac{1}{2}} f(n) = \mu(E) > 0$. Since we know that $f$ is continuous, we can apply the intermediate value theorem and conclude that for some

$a \in (0, \frac{1}{2})$, $0 < f(a) < \mu(E)$. That is,

$$0 < \mu\left(E \cap \left(\mathbf{1}_{\mathcal{F}} \times \left[\frac{1}{2} - a, \frac{1}{2} + a\right]\right)\right) < \mu(E).$$

The event $E \cap (\mathbf{1}_{\mathcal{F}} \times [\frac{1}{2} - a, \frac{1}{2} + a])$ is a subset of $E$ and has a strictly lower measure. Since $E$ was arbitrary, $\langle S \times [0,1], \Sigma, \mu\rangle$ is atomless. □

The following theorem gives a positive answer to problem 4.[1]

**Theorem 10.** *Every probability space can be extended to a probability space which is causally closed w.r.t. $L_{ind}$.*

*Proof.* Let $\langle S, \mathcal{F}, p\rangle$ be a probability space. From lemma 21 we know that $\langle S\times[0,1], \Sigma, \mu\rangle$, which is the product of $\langle S, \mathcal{F}, p\rangle$ with the space of all Borel subsets of the $[0,1]$ segment with the Lebesgue measure, is atomless. Let $h : \mathcal{F} \to \Sigma$ be defined as $h(D) = D\times[0,1]$. It is immediate that $h$ is a Boolean algebra embedding of $\mathcal{F}$ into $\Sigma$, and so $\langle S\times[0,1], \Sigma, \mu\rangle$ is an extension of $\langle S, \mathcal{F}, p\rangle$. Moreover, from fact 18 we infer that it is an extension of $\langle S, \mathcal{F}, p\rangle$ which is *causally closed* w.r.t. $L_{ind}$. □

Regarding the atomless probability spaces, we can in fact prove more: it is not only that the spaces provide for each correlation a statistical common cause; if we confine our attention to correlations between events which are logically independent modulo measure zero events they even provide in each case a countably infinite statistical common cause system (uncountably many such systems, to be precise).

**Theorem 11.** *(Marczyk and Wroński (2014)) Let $\mathbb{S} = \langle \Omega, \mathcal{F}, P\rangle$ be an atomless classical probability space.*

*Then each pair of correlated events from $\mathbb{S}$ belonging to $L^{+}_{ind}$:*

- *has uncountably many countably infinite statistical common cause systems;*
- *has, for any $n \in \mathbb{N}$, $n > 2$, uncountably many statistical common cause systems of size $n$.*

For the lengthy proof of this theorem I refer the Reader to Marczyk and Wroński (2014); an informal tip I can give here is that the main tool is the technique of creating infinite statistical common cause systems from Wroński and Marczyk (2010) (see chapter 2), appropriately amplified either to "stop" at a certain point (thus producing an SCCS of some finite size $n$) or to produce an infinite SCCS lying "exactly where we want it to be" with respect to the correlated events.

1 Theorem 4.2 from Gyenis and Rédei (2011) is to the same effect as my Theorem 10; it has a different proof which uses a different homomorphism. Theorem 10 was a part of my 2010 PhD thesis; Gyenis and Rédei's more complicated technique is to be applied in more general contexts.

The above results may have already convinced the Reader that for those who seek a Reichenbach-style explanation of correlations atomless probability spaces form a veritable paradise, whether one looks for SCCs or SCCSs of any particular size. In the case of SCCs it is possible to show even more. Suppose in some probability space with the measure $P$ events $A$ and $B$ are correlated and $C$ is their statistical common cause. The quintuple $\langle P(C), P(A|C), P(A|C^{\perp}), P(B|C), P(B|C^{\perp})\rangle$ is called the *type* of $C$; this notion was introduced in Hofer-Szabó et al. (1999) where the Authors also derive, assuming that $A$ and $B$ are correlated events, the conditions of admissibility for a quintuple of real numbers to serve as appropriate conditional probabilities for some SCC of $A$ and $B$; whether such an SCC exists in the given space is another matter of course. The main idea is that a single correlation can possess SCCs of various types, which invites a new notion of causal closedness and completability:

**Definition 29**(Rédei[2]) A probability space is strongly causally closed w.r.t. a relation of logical independence $R_{ind}$ if for any correlated pair in $R_{ind}$, and for any type $T$ admissible for that pair, it contains an SCC of type $T$.

A probability space is strongly causally completable w.r.t. $R_{ind}$ if it can be extended to a space which is strongly causally closed w.r.t. $R_{ind}$.

A natural question is this: are any spaces strongly causally closed? If so, are any spaces (which are not strongly causally closed themselves) strongly causally completable? The answer is "yes" on both counts, and I guess at this point we should not be surprised as to which spaces are doing the job:

**Theorem 12.** *(Marczyk and Wroński (2013)) All atomless classical probability spaces are strongly causally closed w.r.t. $L^{+}_{ind}$. Since all classical probability spaces can be extended to atomless classical probability spaces, all classical probability spaces are also strongly causally completable.*

Proof sketch:

- Take an atomless probability space with measure $P$, two correlated events $A$ and $B$ in it, and a type $T$ admissible for that correlation;[3]
- the type $T$ is given by the probabilities $P(C), P(A|C), P(A|C^{\perp}), P(B|C), P(B|C^{\perp})$ for a common cause $C$ of that type;
- this in turn determines the probabilities $P(CAB)$, $P(CA^{\perp}B)$, $P(CAB^{\perp})$, and $P(CA^{\perp}B^{\perp})$ a common cause $C$ of type $T$ for $A$ and $B$ would have;

2 This definition and the following problem have been presented by M. Rédei during the "New Directions in the Philosophy of Science" conference in Bertinoro, October $17^{th}$-$20^{th}$ 2012.

3 Hofer-Szabó et al. (1999) give a detailed definition of admissibility.

- we can cut out four "chunks" inside $AB$, $AB^{\perp}$, $A^{\perp}B$, and $A^{\perp}B^{\perp}$ (which exist because the space is atomless) and take their union to be $C$, a common cause of type $T$ for the original correlated pair.

It would seem natural to extend the notion of type so that it would pertain not only to SCCs but also to SCCSs; we plan to approach this task in Marczyk and Wroński (2013), and I will refrain from doing so here, since I believe the main point about atomless classical probability spaces has been made: in short, any classical space can be extended to an atomless one and when it comes to Reichenbach-style explanations of correlations, *everything is possible* in atomless spaces.

## 6.3 Some Remarks on Causal Completability of Non-classical Probability Spaces

Another result from Hofer-Szabó et al. (1999) says that every nonclassical space with an additive state $\mu$ has an extension containing SCCs for every pair correlated in $\mu$:

**Fact 22** (Proposition 3 from Hofer-Szabó et al. (1999)). *Let $\langle S, \mu\rangle$ be a non-classical probability space. Let $\mathcal{F}$ be the family of all pairs of events correlated in $\mu$. Then $\langle S, \mu\rangle$ is causally completable w.r.t. the family $\mathcal{F}$.*

However, one could ask a question similar to the one answered in the previous section: is any non-classical probability space extendable to a causally closed non-classical probability space? So far, the answer to this question is not known; for some of the latest developments see Gyenis and Rédei (2011).

One way of approaching the problem would be to try to use Kitajima's result presented above as fact 19. The task would then boil down to the following: given a non-classical space $\langle L, P\rangle$, find its extension $\langle L', P'\rangle$ with an atomless $L'$. The extension would have to be an atomless non-classical probability space and would be, by fact 19, causally closed w.r.t. $L_{ind}$.

Unfortunately, it seems we should not count on using a method similar to the one outlined in the previous section. The strategy was this: take a Boolean algebra and construct a product of it and an (atomic) Boolean algebra being the event space of an atomless probability space. The crucial point is that there is a subalgebra in the product which is isomorphic to the original algebra. Thus we can find a homomorphism due to which the product can serve as the event space for an extension of the original space.

Algebraically speaking, the above fact can be phrased as "for any Boolean algebras $A$ and $B$, $A$ is a retract of $A \times B$" (see p. 90 of Koppelberg (1989)). However, once we switch to non-distributive lattices, we can no longer count on this. If $A$ is a non-distributive OML and $B$ is a Boolean algebra, it may happen that the product $A \times B$

does not contain a subalgebra isomorphic to $A$. Thus such a product in general does not seem to be a good candidate for the OML on which an additive state could be defined so that the resulting non-classical probability space would be an extension of the initial space.

To see this, take the product of the so called "Chinese lantern" (on the left) and the two-element Boolean algebra:

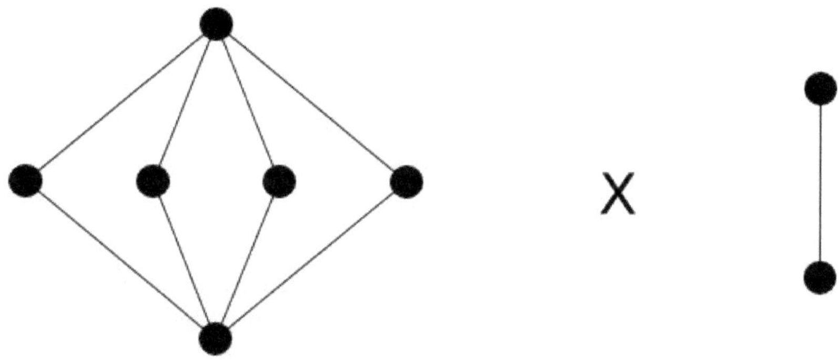

which is the following orthomodular lattice:

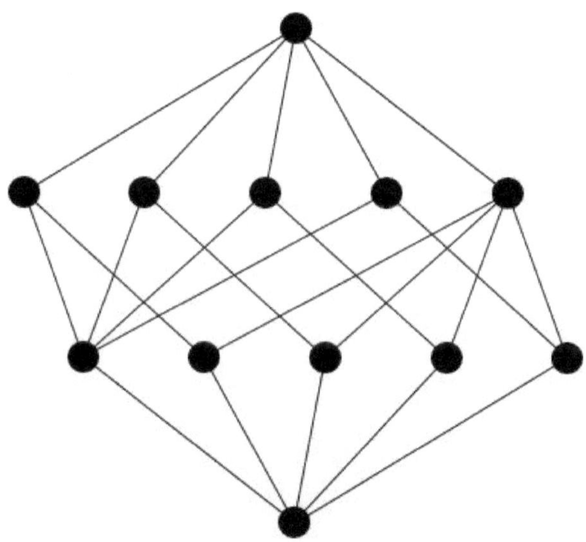

The orthocomplement function has not been displayed on the pictures, but it can be easily checked that the above lattice does not have a subalgebra isomorphic to the Chinese lantern for any proper ascription of orthocomplements.[4]

Therefore, when trying to construct for a given OML $L$ an OML $L'$ which would contain a subalgebra isomorphic to $L$, taking a product of $L$ and some other OML may be a bad move.[5] This is even before we take into account the fact that we would like to require that $L'$ be atomless and complete. In fact, the general issue whether every OML can be embedded into a complete OML is a long-standing open problem (see e.g. chapter 8 in Bruns and Harding (2000)).

## 6.4 The Impact of the Results

In light of the completability results I presented it seems that, in the context of classical probability spaces, PCC 4 is a true principle. Of course, it is not very practical. To make a brief foray into decidedly non-formal matters, perhaps a god or some other vastly knowledgeable entity could envisage an uncountable probability space which takes into account every possible factor and contains SCCs for all possible correlations of logically independent events. The role of PCC 4 would be that of reassurance of us puny humans.

At any rate, the upshot is that to disprove the PCC it is never enough to point at a family of correlations lacking a Reichenbach-style explanation in a certain probability space; it is *always* possible to extend the space into a space containing Reichenbach-style explanations, seemingly regardless of the restrictions we may have put on them (we could demand some specific cardinality or some specific types of SCCs). On the one hand, in a sense this shows that the PCC is not falsifiable, since if seemingly falsifying evidence was proposed one could answer by producing a probability space causally closed in the desired sense, which would be compatible with that evidence. On the other hand, since it is so "easy" to find Reichenbach-style explanations of correlations, Reichenbach's notion of a statistical common cause and its descendant notions on their own do not seem to form a viable basis for a philosophical project revolving around the analysis of causality. Perhaps they could be used to form an interesting addition to such a project but they seem to share the fate of all that is ubiquitous: it is usually also not too interesting on its own. (I would take the fact that they *are* ubiquitous to be interesting and surprising, though.)

4 The point could perhaps be made clearer if we restricted our attention just to the two lattice operations $\vee$ and $\wedge$: the last OML represented not only does not have a subalgebra isomorphic to the Chinese lantern taken as an algebra with three operations, but (which is immediately seen) does not even have a sublattice isomorphic to the Chinese lantern taken as a lattice.

5 Of course, our example presented a quite special case, since the "other" OML in it was Boolean.

# Conclusion

The main theorems of this study are included in chapters 5 and 6. Among them are positive results concerning the prospects of explaining correlations *via* statistical common cause systems. Perhaps of the biggest intuitive force is theorem 10: every probability space can be extended to a probability space which is causally closed. (Subsequent theorems express variations on this idea.) If we do not see a statistical common cause for two correlated events, it is only because we have directed our attention to the "wrong" probability space; there is an extension of it which leaves no correlations unexplained. However, as already mentioned in the introduction to this essay, this can be interpreted in two ways. On the one hand, it is always nice to have a general positive theorem about the applicability of some interesting notion. On the other hand, in this case the applicability may be strictly mathematical. It is by no means evident that the statistical common causes present in the "extended" spaces will have much to do with what we would naturally accept as causes.

Consider again the example of particle decay from section 1.4.1. The momentum of one part of the particle is determined in accordance with the principle of conservation of total momentum by the momentum of the other particle part. Suppose the experimental setup is described by a probability space $\mathcal{S}$. The state of the particle before the decay event is, in this space, not a screener-off for the values of momentum after the split. Consider then the causally closed space $\mathcal{S}'$, which is an extension of $\mathcal{S}$, and which has to exist by theorem 10. In this space the momenta of the two parts of the decayed particle possess a statistical common cause which screens off one from the other. But do we really expect such a screener to have anything to do with the true causal picture? Are we not fully satisfied with the explanation consisting of the principle of conservation of total momentum coupled with the information on the genesis of the two particle parts?

Due to the generality of theorem 10 it is true that the "big space" used in the eponymous approach to the Bell inequalities can also be extended to a causally closed space—in which all correlations have common causes. This may be surprising if one thought that this should lead to the empirically falsified inequalities. This is, however, not the case, since—as already noted–in the "big space" approach the EPR-type correlations are in fact *conditional* correlations and so do not fall under the scope of theorem 10. And if we move to the "many spaces" approach then each of the "small" spaces will have their own causally closed extension. However, it is by no means evident that Parameter Independence, Outcome Independence and No Conspiracy should hold for the "extended" spaces.

In conclusion, we should better be skeptical towards a general application of explaining correlations by means of purely probabilistically-defined notions. Those employed in this essay, from Reichenbach's common cause as the middle element of a conjunctive fork, through statistical common cause systems, to *deductive explanantes*, share (apart from screening off considered in isolation) the deductive explanatory feature described in section 2.1. This is a pleasing fact which strengthens the case for such notions playing a role in explanation but it is clearly not enough, e.g. since the correlation itself—as well as many other more or less trivially equivalent (sets of) conditions—also has that particular feature. There is of course more to causal explanation than pure statistics. However, mathematical methods like the algorithm of constructing SCCs and SCCSs used in the proof of theorem 4 could perhaps provide *candidates* for explanations—for example, they may suggest searching for traits possessed by certain subsets of the examined population—which can subsequently be studied by applying other methods: using the previous knowledge of mechanisms operating in the given context. This is at least *a priori* possible but if someone is of the opinion that the prospects are rather dim I will not try to convince them otherwise.

If the Reader will allow me a tongue-in-cheek remark, I think we could say that in light of the results of the previous chapters the atomless probability spaces offer us something akin to "Reichenbach's paradise": a realm in which all correlations are explained by factors satisfying Reichenbach's conditions. Now the task would be to find a guide through that paradise, a—presumably non-mathematical—tool for staying on the road of real causal explanation and steering away from bogus, non-explanatory constructions.

This is because the various results in this book suggest the possibility of upholding the PCC "against all odds", by pointing to SCCs (or SCCSs etc.) of the supposedly unexplained correlations in an empirically adequate probability space. The chief worry is that cause-wise and explanation-wise these might simply be "false positives"—events (or constructions made of events) satisfying the probabilistic requirements but not corresponding to anything causally and / or explanatorily relevant. But since the theorems already employ a notion of independence, which invites the use of background causal knowledge in order to exclude some pairs of logically independent events from the considerations (since correlations between them are not surprising, because e.g. we already know one is a cause of the other), why not invite the use of similar knowledge again, this time on the level of Reichenbach-style explanations of correlations? The results in this book could perhaps be seen as usable as a mathematical "bridge" between two applications of background causal knowledge:
- first, the background causal knowledge determines which correlations between logically independent events demand explanation;
- then the theorems from chapters 5 and 6 point to candidates for such explanation;
- then the background causal knowledge allows us to "weed out" the obviously irrelevant among those explanations.

At this point I am unfortunately unable to say anything specific on this issue. It is not clear to me whether any specific picture of causality would fit better with such a view of the PCC than any other. Perhaps some pluralistic account of causality and explanation of correlations is possible, one which would combine a probabilistic analysis of some aspects of causality with e.g. a mechanistic (or some other) analysis of different aspects. Whether such an approach is indeed possible would be a topic for another book. I would very much like to read such a book some day but right now I would not even be able to begin writing it. This is the time to stop, then.

# Bibliography


Adams, E. W. (1998). *A Primer of Probability Logic.* CSLI Publications (Univ. of Chicago Press).

Arntzenius, F. (1992). The Common Cause Principle. *PSA: Proceedings of the Biennial Meeting of the Philosophy of Science Association*, Vol. 1992, Volume Two: Symposia and Invited Papers.

Arntzenius, F. (2005). Reichenbach's Common Cause Principle. In Zalta, E., editor, *The Stanford Encyclopedia of Philosophy (Winter 2008 Edition).* http://plato.stanford.edu/archives/win2008/entries/physics-Rpcc/.

Aspect, A., Dalibard, J., and Gérard, R. (1982). Experimental test of Bell's Inequalities using time-varying analyzers. *Physical Review Letters*, 49:1804–1807.

Bell, J. S. (1964). On the Einstein–Podolsky–Rosen Paradox. *Physics*, 1:195–200.

Belnap, N. and Szabó, L. E. (1996). Branching Space-Time Analysis of the GHZ Theorem. *Foundations of Physics*, 26(8):989–1002.

Berkovitz, J. (2000). The Many Principles of the Common Cause. *Reports on Philosophy*, 20.

Berkovitz, J. (2008). Action at a Distance in Quantum Mechanics. In Zalta, E., editor, *The Stanford Encyclopedia of Philosophy (Winter 2008 Edition).* http://plato.stanford.edu/archives/win2008/entries/qm-action-distance/.

Billingsley, P. (1995). *Probability and Measure.* John Wiley & Sons, Inc., 3rd edition.

Blalock, H. (1979). *Social Statistics.* McGraw-Hill, 2nd edition.

Bruns, G. and Harding, J. (2000). Algebraic Aspects of Othomodular Lattices. In Coecke, B., Moore, D., and Wilce, A., editors, *Current research in operational quantum logic: algebras, categories, languages*, pp. 37–65. Springer.

Butterfield, J. (1989). A Space-Time Approach to the Bell Inequality. In Cushing, J. T. and McMullin, E., editors, *Philosophical Consequences of Quantum Theory. Reflections on Bell's Theorem.* University of Notre Dame Press, Indiana.

Butterfield, J. (2007). Stochastic Einstein Locality Revisited. *The British Journal for the Philosophy of Science*, 58(4):805–867.

Cartwright, N. (1988). How to tell a common cause: generalizations of the conjunctive fork criterion. In Fetzer, J. H., editor, *Probability and Causality*, pp. 181–188. D. Reidel Publishing Company.

Cartwright, N. (1989). *Nature's Capacities and Their Measurement.* Oxford University Press.

Cartwright, N. (1999). *The Dappled World. A Study of the Boundaries of Science.* Cambridge University Press.

Cartwright, N. (2002). Against modularity, the Causal Markov Condition, and any link between the two: comments on Hausman and Woodward. *The British Journal for the Philosophy of Science*, 53(3):411–453.

Clauser, J. F. and Horne, M. A. (1974). Experimental consequences of objective local theories. *Physical Review D*, 10(2):526–535.

Eberhardt, F. (2011). Reliability via synthetic a priori: Reichenbach's doctoral thesis on probability. *Synthese*, 181:125-136.

Feller, W. (1968). *An Introduction to Probability Theory and its Applications.* John Wiley & Sons, Inc.

Fine, A. (1982a). Hidden Variables, Joint Probability, and the Bell Inequalities. *Physical Review Letters*, 48(5):291–295.

Fine, A. (1982b). Some local models for correlation experiments. *Synthese*, 50:279–294.

Forster, M. R. (1988). Sober's Principle of Common Cause and the Problem of Comparing Incomplete Hypotheses. *Philosophy of Science*, 55(4):538–559.

Fremlin, D. (2001). *Measure Theory*, volume 2. Torres Fremlin.

Glymour, C. (2010). What Is Right with 'Bayes Net Methods' and What Is Wrong with 'Hunting Causes and Using Them'? *The British Journal for the Philosophy of Science*, 61:161–211.

Graßhoff, G., Portmann, S., and Wüthrich, A. (2005). Minimal Assumption Derivation of a Bell-type Inequality. *The British Journal for the Philosophy of Science*, 56:663–680.

Gyenis, B. and Rédei, M. (2004). When Can Statistical Theories be Causally Closed? *Foundations of Physics*, 34(9):1284–1303.

Gyenis, B. and Rédei, M. (2010). Causal completeness in general probability theories. In Suárez, M., editor, *Probabilities, Causes, and Propensities in Physics*. Synthese Library, Springer.

Gyenis, Z. and Rédei, M. (2011). Characterizing common cause closed probability spaces. *Philosophy of Science*, 78:393–409.

Haig, B. D. (2003). What Is a Spurious Correlation? *Understanding Statistics*, 2(2):125–132.

Hausman, D. M. and Woodward, J. (1999). Independence, invariance and the causal Markov condition. *The British Journal for the Philosophy of Science*, 50(4):521–583.

Hausman, D. M. and Woodward, J. (2004). Modularity and the causal Markov condition: a restatement. *The British Journal for the Philosophy of Science*, 55(1):147–161.

Higashi, K. (2008). The Limits of Common Cause Approach to EPR Correlation. *Foundations of Physics*, 38:591–609.

Hitchcock, C. (2010). Probabilistic Causation. In Zalta, E., editor, *The Stanford Encyclopedia of Philosophy (Summer 2010 Edition)*. http://plato.stanford.edu/archives/sum2010/entries/causation-probabilistic/.

Hofer-Szabó, G. (2008). Separate- versus common-common-cause-type derivations of the Bell inequalities. *Synthese*, 163:199–215.

Hofer-Szabó, G. (2011). Bell($\delta$) inequalities derived from separate common causal explanation of almost perfect EPR anticorrelations. *Foundations of Physics*, 41:1398–1413.

Hofer-Szabó, G. (2012). Separate common causal explanation and the Bell inequalities. *International Journal of Theoretical Physics*, 51:110–123.

Hofer-Szabó, G. and Rédei, M. (2004). Reichenbachian Common Cause Systems. *International Journal of Theoretical Physics*, 43(7/8):1819–1826.

Hofer-Szabó, G. and Rédei, M. (2006). Reichenbachian Common Cause Systems of arbitrary finite size exist. *Foundations of Physics*, 36(5):745–756.

Hofer-Szabó, G., Rédei, M., and Szabó, L. E. (1999). On Reichenbach's common cause principle and Reichenbach's notion of common cause. *The British Journal for the Philosophy of Science*, 50(3):377–399.

Hofer-Szabó, G., Rédei, M., and Szabó, L. E. (2000). Reichenbach's Common Cause Principle: Recent Results and Open Questions. *Reports on Philosophy*, 20:85–107.

Hofer-Szabó, G., Rédei, M., and Szabó, L. E. (2002). Common-Causes are Not Common Common-Causes. *Philosophy of Science*, 69:623–636.

Hoover, K. D. (2003). Nonstationary Time Series, Cointegration, and the Principle of the Common Cause. *The British Journal for the Philosophy of Science*, 54:527–551.

Jarrett, J. P. (1984). On the Physical Significance of the Locality Conditions in the Bell Arguments. *Noûs*, 18(4):569–589.

Kalmbach, G. (1983). *Orthomodular Lattices*. Academic Press.

Kitajima, Y. (2008). Reichenbach's Common Cause in an Atomless and Complete Orthomodular Lattice. *International Journal of Theoretical Physics*, 47:511–519.

Koppelberg, S. (1989). *Handbook of Boolean Algebras*, volume 1. North-Holland. (J. Donald Monk and Robert Bonnet, Eds.).

Marczyk, M. and Wroński, L. (2013). Strong causal completability of classical probability spaces. In preparation.

Marczyk, M. and Wroński, L. (2014). Completion of the Causal Completability Problem. *British Journal for the Philosophy of Science*. Forthcoming.

Mill, J. S. (1868). *A System of Logic*. Longmans, Green, Reader, and Dyer.

Müller, T. and Placek, T. (2001). Against a minimalist reading of Bell's theorem: Lessons from Fine. *Synthese*, 128:343–379.

Palm, G. (1978). Additive entropy requires additive measure. *Archiv der Mathematik*, 30(1):293–296.

Pearl, J. (1988). *Probabilistic Reasoning in Intelligent Systems*. San Mateo, CA: Morgan Kaufmann.

Pearl, J. (2000). *Causality. Models, Reasoning, and Inference*. Cambridge University Press.

Placek, T. (2000). *Is Nature Deterministic? A Branching Perspective on EPR Phenomena*. Jagiellonian University Press.

Placek, T. (2009). The Distinction Common Cause vs. Common Common Cause is a Red Herring. Talk given at the EPSA 2009 Conference, Amsterdam, 21-24 X 2009.

Placek, T. (2010). On Propensity-Frequentist Models for Stochastic Phenomena with Applications to Bell's Theorem. In Czarnecki, T., Kijania-Placek, K., Kukushkina, V., and Woleński, J., editors, *The Analytical Way. Proceedings of the 6th European Congress of Analytic Philosophy*, London. College Publications.

Placek, T. and Wroński, L. (2011). Separate common causes and EPR correlations—a no-go Result. Archived at the PhilSci Archive, http://philsci-archive.pitt.edu/8617/.

Portmann, S. and Wüthrich, A. (2007). Minimal assumption derivation of a weak Clauser–Horne inequality. *Studies in History and Philosophy of Science Part B: Studies in History and Philosophy of Modern Physics*, 38(4).

R Core Team (2012). *R: A Language and Environment for Statistical Computing*. R Foundation for Statistical Computing, Vienna, Austria. ISBN 3-900051-07-0.

Reichenbach, H. (1949). *The theory of probability*. University of California Press.

Reichenbach, H. (1971). *The Direction of Time*. University of California Press. Reprint of the 1956 edition.

Reiss, J. (2007). Time Series, Nonsense Correlations and the Principle of the Common Cause. In Russo, F. and Williamson, J., editors, *Causality and Probability in the Sciences*. College Publications.

Russell, B. (2009). *Human Knowledge: Its Scope and Limits*. Routledge Classics. Taylor & Francis Routledge, London and New York. Reprint of the 1948 edition.

Salmon, W. C. (1971). *Statistical Explanation and Statistical Relevance*. University of Pittsburgh Press. (With contributions by Richard C. Jeffrey and James G. Greeno).

Salmon, W. C. (1984). *Scientific Explanation and the Causal Structure of the World*. Princeton University Press.

Salmon, W. C. (1998a). Causality without Counterfactuals. In *Causality and Explanation*, pp. 248–260. Oxford University Press.

Salmon, W. C. (1998b). Why ask, 'Why?'? An Inquiry Concerning Scientific Explanation. In *Causality and Explanation*, pp. 125–141. Oxford University Press.

Scheidl, T., Ursin, R., Kofler, J., Ramelow, S., Ma, X.-S., Herbst, T., Ratschbacher, L., Fedrizzi, A., Langford, N., Jennewein, T., and Zeilinger, A. (2010). Violation of local realism with freedom of choice. *Proceedings of the National Academy of Sciences of the United States of America*, 107(46):19708-19713.

Shimony, A. (2009). Bell's Theorem. In Zalta, E., editor, *The Stanford Encyclopedia of Philosophy (Summer 2009 Edition)*. http://plato.stanford.edu/archives/sum2009/entries/bell-theorem/.

Sober, E. (1988). The Principle of the Common Cause. In Fetzer, J. H., editor, *Probability and Causality*, pp. 211–228. D. Reidel Publishing Company.

Sober, E. (1992). Screening-off and the units of selection. *Philosophy of Science*, 59:142–152.

Sober, E. (2001). Venetian Sea Levels, British Bread Prices, and the Principle of the Common Cause. *The British Journal for the Philosophy of Science*, 52:331–346.

Spirtes, P., Glymour, C., and Scheines, R. (2000). *Causation, Prediction, and Search*. The MIT Press, 2nd edition.

Steel, D. (2005). Indeterminism and the Causal Markov Condition. *The British Journal for the Philosophy of Science*, 56:3–26.

Steel, D. (2006). Comment on Hausman & Woodward on the Causal Markov Condition. *The British Journal for the Philosophy of Science*, 57(1):219–231.

Suárez, M. (2007). Causal inference in quantum mechanics: a reassessment. In Russo, F. and Williamson, J., editors, *Causality and Probability in the Sciences*. College Publications.

Szabó, L. E. (2000). Attempt to resolve the EPR–Bell paradox via Reichenbachian concept of common cause. *International Journal of Theoretical Physics*, 39(3):901–911.

Szabó, L. E. and Fine, A. (2002). A local hidden variable theory for the GHZ experiment. *Physics Letters A*, 295:229–240.

Torretti, R. (1987). Do Conjunctive Forks Always Point to a Common Cause? *The British Journal for the Philosophy of Science*, 38(3):384–387.

Uffink, J. (1999). The Principle of the Common Cause Faces the Bernstein Paradox. *Philosophy of Science*, 66(Supplement. Proceedings of the 1998 Biennial Meetings of the Philosophy of Science Association. Part I: Contributed Papers):S512–S525.

van Fraassen, B. C. (1980). *The Scientific Image*. Oxford, Clarendon Press.

van Fraassen, B. C. (1982). The Charybdis of Realism: Epistemological Implications of Bell's Inequality. *Synthese*, 52:25–38. Reprinted with additions in Cushing, JT., McMullin, E. (eds) (1989).

van Fraassen, B. C. (1991). *Quantum Mechanics. An Empiricist View*. Oxford, Clarendon Press.

Verma, T. and Pearl, J. (1988). Causal networks: Semantics and expressiveness. In *Proceedings of the 4th Workshop on Uncertainty in Artificial Intelligence (Mountain View, CA)*, pp. 352–359. Reprinted in R. Schachter, T. S. Levitt, and L. N. Kanal (Eds.), Uncertainty in Artificial Intelligence, vol. 4, pp. 69-76.

Williamson, J. (2005). *Bayesian Nets and Causality. Philosophical and Computational Foundations*. Oxford University Press.

Williamson, J. (2009). Probabilistic theories of causality. In Beebee, H., Hitchcock, C., and Menzies, P., editors, *The Oxford Handbook of Causation*, pp. 185–212. Oxford University Press.

Woodward, J. (2005). *Making things happen: A theory of causal explanation*. Oxford University Press.

Wroński, L. (2010). *The Common Cause Principle. Explanation via screening off*. PhD thesis, Jagiellonian University, Kraków. Archived at http://jagiellonian.academia.edu/LeszekWroński.

Wroński, L. and Marczyk, M. (2010). Only Countable Reichenbachian Common Cause Systems Exist. *Foundations of Physics*, 40:1155–1160.

Wroński, L. and Marczyk, M. (2013). A New Notion of Causal Closedness. *Erkenntnis*. Forthcoming, DOI 10.1007/s10670-013-9457-0.

# Index

Bei Fragen zur Produktsicherheit wenden Sie sich bitte an:
If you have any questions regarding product safety,
please contact:

Walter de Gruyter GmbH
Genthiner Straße 13
10785 Berlin
productsafety@degruyterbrill.com